W0106580

VHDL'92

by

Jean-Michel Bergé
CNET, France Telecom, France
Alain Fonkoua
Institut Méditerranéen de Technologie, France
Serge Maginot
LEDA S.A., France
Jacques Rouillard
Institut Méditerranéen de Technologie, France

Springer Science+Business Media, LLC

ISBN 978-1-4613-6427-6 ISBN 978-1-4615-3246-0 (eBook)
DOI 10.1007/978-1-4615-3246-0
Library of Congress Cataloging-in-Publication Data

A C.I.P. Catalogue record for this book is available from
the Library of Congress.

Printed on acid-free paper.

VHDL'92

THE KLUWER INTERNATIONAL SERIES IN ENGINEERING AND COMPUTER SCIENCE

VLSI, COMPUTER ARCHITECTURE AND DIGITAL SIGNAL PROCESSING

Consulting Editor
Jonathan Allen

Other books in the series:

HOT-CARRIER RELIABILITY OF MOS VLSI CIRCUITS, Y. Leblebici, S. Kang
ISBN: 0-7923-9352-X
MOTION ANALYSIS AND IMAGE SEQUENCE PROCESSING, M. I. Sezan, R. Lagendijk
ISBN: 0-7923-9329-5
HIGH-LEVEL SYNTHESIS FOR REAL-TIME DIGITAL SIGNAL PROCESSING: The Cathedral-II Silicon Compiler, J. Vanhoof, K. van Rompaey, I. Bolsens, G. Gossens, H. DeMan
ISBN: 0-7923-9313-9
SIGMA DELTA MODULATORS: Nonlinear Decoding Algorithms and Stability Analysis, S. Hein, A. Zakhor
ISBN: 0-7923-9309-0
LOGIC SYNTHESIS AND OPTIMIZATION, T. Sasao
ISBN: 0-7923-9308-2
ACOUSTICAL AND ENVIRONMENTAL ROBUSTNESS IN AUTOMATIC SPEECH RECOGNITION, A. Acero
ISBN: 0-7923-9284-1
DESIGN AUTOMATION FOR TIMING-DRIVEN LAYOUT SYNTHESIS, S. S. Sapatnekar, S. Kang
ISBN: 0-7923-9281-7
DIGITAL BiCMOS INTEGRATED CIRCUIT DESIGN, S. H. K. Embadi, A. Bellaouar, M. I. Elmasry
ISBN: 0-7923-9276-0
WAVELET THEORY AND ITS APPLICATIONS, R. K. Young
ISBN: 0-7923-9271-X
VHDL FOR SIMULATION, SYNTHESIS AND FORMAL PROOFS OF HARDWARE, J. Mermet
ISBN: 0-7923-9253-1
ELECTRONIC CAD FRAMEWORKS, T. J. Barnes, D. Harrison, A. R. Newton, R. L. Spickelmier
ISBN: 0-7923-9252-3
ANATOMY OF A SILICON COMPILER, R. W. Brodersen
ISBN: 0-7923-9249-3
FIELD-PROGRAMMABLE GATE ARRAYS, S. D. Brown, R. J. Francis, J. Rose, S. G. Vranesic
ISBN: 0-7923-9248-5
THE SECD MICROPROCESSOR, A VERIFICATION CASE STUDY, B. T. Graham
ISBN: 0-7923-9245-0
HIGH LEVEL SYNTHESIS OF ASICs UNDER TIMING AND SYNCHRONIZATION CONSTRAINTS, D. C. Ku, G. De Micheli
ISBN: 0-7923-9244-2
FAULT COVERING PROBLEMS IN RECONFIGURABLE VLSI SYSTEMS, R. Libeskind-Hadas, N. Hassan, J. Cong, P. McKinley, C. L. Liu
ISBN: 0-7923-9231-0
VHDL DESIGNER'S REFERENCE, J-M. Bergé, A. Fonkoua, S. Maginot, J. Rouillard
ISBN: 0-7923-1756-4

«Qu'il est beau, ton dessin !»
dit la maman au bébé couvert de peinture.

«J'ai toujours rêvé qu'on me dédie un livre sur VHDL»,
nous diront nos femmes, compagnes, parents, enfants.

Voilà qui est fait.

TABLE OF CONTENTS

Introduction

New Simulation Mechanisms

New Structuring Mechanisms

New Interfacing Mechanisms

New Predefined Operators, Functions & Attributes

Slight Enhancements

Language Simplifications

Clarifications

Annex

LIST OF FIGURES

FOREWORD

This book is not, in any respect, a reference manual for the VHDL language. That purpose is served by the VHDL Language Reference Manual (LRM), which defines the language in a formal, contractual, and unambiguous manner. Each feature of the language is carefully described in the LRM, but the reader of that manual is expected to be a tool builder or an experienced designer.

The present book gives an introduction to new VHDL language constructs and illustrates some of their possible uses. The reader should be able to use the construct and to understand the LRM where these new features are defined. This will allow him or her to locate all the already existing details of the language that are outside the scope of this book.

Reading this book from beginning to end is certainly an effective way to reach a better understanding of the new possibilities of VHDL'92. Nevertheless, since VHDL is not only a hardware description language but also an environment, the reader may be more interested in a particular topic or domain of application. Chapter 1 lists many of these concerns and can guide the reader to the more pertinent features in his or her domain.

The terms VHDL'87 and VHDL'92 as used in this book need preliminary explanation. VHDL'92 is the IEEE standard 1076, balloted in December 1987. Since VHDL'92 is a new version of standard 1076, this vote renders the previous standard (IEEE 1076/87) obsolete. Basically, there will not be two standards (VHDL'87 and VHDL'92), but only one. However, since the purpose of this book is to present the new features to people aware of IEEE 1076/87, we will nevertheless refer to this old version as VHDL'87.

Acknowledgments

This book has been made possible thanks to:

• The European ESPRIT/ECIP2 project in which the CNET[1] and the IMT[2] are involved.

• The sponsorship of ECIP by CEC[3], and of VHDL standardization activities by France Telecom and LEDA[4].

• The support the authors received from their management, Pierre Doucelance, Roland Gerber, Jean-Louis Lardy, Jacques Lecourvoisier, Christian Maresca, Jean Mermet, Jean Pierre Noblanc, Joël Rodriguez, from Robin Lafontaine who runs the ECIP project and from Mike Newman of the CEC.

• The fruitful preliminary review of Roland Airiau, Vincent Olive, Jacques Pulou, Anne Robert, Denis Rouquier.

• The very useful comments given by Maureen Timmins, by the anonymous reviewers provided by Kluwer Academic Publishers, and by Mike Casey, editor at Kluwer.

• The excellent copy-editing provided by Gerry Geer.

The authors would also like to highlight that all the IEEE standardization work, supported by companies and institutions in terms of traveling and time, has been carried out by individuals under their own responsibility; that many meetings and workshops have been held at weekends, many emails have been dated after midnight. We acknowledge all our friends, colleagues and relations who are already listed in the foreword of the LRM, and their families who are not.

---

[1] Centre National d'Etudes des Télécommunications (France Telecom), BP 42 38240 Meylan Cedex, France.

[2] Institut Méditerranéen de Technologie, 13451 Marseille Cedex 20, France.

[3] Commission of the European Community, 200 rue de la Loi, Brussels, Belgium.

[4] Languages for Design Automation, 35 av du Granier, 38240 Meylan, France.

=> ***Introduction***

New Simulation Mechanisms
New Structuring Mechanisms
New Interfacing Mechanisms
New Predefined Operators, Functions & Attributes
Slight Enhancements
Language Simplifications
Clarifications
Annex

1. DESIGNER'S CONCERNS

Since it is difficult to find a logical order for classifying application domains such as synthesis or system design, or pragmatic considerations such as portability or compatibility with VHDL'87, the following sections are presented in alphabetical order.

1.1. ANALOG(MIXED-MODE SIMULATION)

Among all the requests from users for a VHDL'92 standardization, the requirements concerning analog domain, and more precisely mixed-mode simulation, were most numerous: about one fifth of the total. Europeans are particularly interested in this field.

In the very early stages of the standardization process, it became apparent that it would not be possible, due to the very different levels of requests as well as to scheduling, for analog extension to be included in VHDL'92. However, because of the interest of designers in this topic, a PAR (a Project Authorization Request) of VHDL has been launched. It is expected to be balloted at the end of 1993 or at the beginning of 1994. The main characteristic of this PAR is that it will be an overset of the language, that includes VHDL'92 as the digital part.

Thus, VHDL'92 (like VHDL'87) does not cover the analog domain. Nevertheless, it is easy to find many papers and even tutorials explaining how to model analog parts in VHDL'87. For limited purposes, this has been done with VHDL'87 and, of course, is possible with VHDL'92.

Indeed, VHDL'92 offers some new possibilities for this purpose. The main one is the introduction of a foreign "mechanism" (see chapter 8) that allows foreign models to be interfaced. These models can potentially be described in an analog modeling language. This is not a "pure" VHDL solution, and the interface is described so briefly that problems of portability and dependency on tool vendors are inevitable. Nevertheless, this interface does exist, which is an important advantage.

For people in search of a "pure" VHDL solution, one of the main problems is to identify the "steady states" of the digital world during which the analog kernel may execute. Last-delta activation processes (see chapter 3) should solve this. It is also possible to imagine calling foreign subprograms (chapter 8) from these processes to activate an external analog kernel.

Analog modeling very often depends on parameters that can vary dynamically during simulation. One example of this is temperature. The dynamic characteristics of such parameters cannot be represented by the use of generic parameters. Their representation as global signals or specific ports is definitely not satisfying. Shared variables (see chapter 4), which are in fact global variables, have been introduced into VHDL'92 and are adapted to this description.

1.2. BACKANNOTATION

A global mechanism for backannotation has not been introduced into VHDL'92. Despite much effort, all the studies in this domain result either in specific solutions that do not cover all possible cases, or in flexible and complex solutions that break the consistency of the language. Furthermore, a consensus has never been reached on the nature of the problem itself: must the backannotation mechanism be included in the language, or is it an external tool, like a debugger ? (We do not handle breakpoints in the language.)

In the past, some papers [MAG] have shown how to build the backannotation mechanism into the language itself. Although not very efficient in terms of performance, and very often cumbersome to write, these solutions work. Certain VHDL'92 features contribute to their implementation.

Incremental configuration (see chapter 6) provides a flexible way to delay the binding of the generics related to backannotation until very late in the elaboration phase.

It can be useful, from within an architecture body, to know which instantiation is modeled. This information was only obtainable in VHDL by using an extra generic parameter: the instantiation identity. A predefined attribute PATH_NAME (chapter 17) returns a string including this information.

Backannotation mechanisms usually read the timing values from a file during elaboration (i.e., before simulation, but when the network is known). Most of the VHDL'87 implementations allow objects to be initialized by calling a function reading a file. The VHDL'87 LRM is unclear on this point. This feature is explicitly possible in VHDL'92 and therefore portable.

[MAG] Electronique, May 92 - Intégrer VHDL dans un environnement de conception d'ASIC - Serge Maginot

It should also be noted that a standardization group defines a format to store backannotation information in a file. Whatever the tool may be, VHDL written or external, this format, when adopted, should be taken into account.

1.3. COMPATIBILITY PROBLEMS VERSUS VHDL'87

Ensuring upward compatibility has been a serious concern of the language design team. Indeed, for obvious industrial reasons, accomplishing this goal was the first guideline of the standardization process.

Although if, strictly speaking, VHDL'92 is not fully compatible, the changes to be made on a VHDL'87 source code to produce a legal VHDL'92 code are few and simple.

As shown in the list of reserved words (chapter 37), some new reserved words have been added. Therefore, a VHDL'87 identifier whose name is one of the new reserved words is now illegal. The chapter discussing identifier generalization syntax (see chapter 22) proposes a strategy for the renaming of identifiers without loss of readability.

The attributes 'STRUCTURE and 'BEHAVIOR have been removed from the language. This is a potential source of non-compatibility, but a quick survey of people involved in the standardization process has shown that, due to the poor definitions of these attributes, nobody actually used them.

The extension of the character set, type CHARACTER of the package STANDARD, is described in chapter 21. The extension can lead to incompatibility when using some predefined attributes on this type.

The semantics of the concatenation operator (&) has been changed according to the index subtype of the result base type (see chapter 28). Potential problems of compatibility may occur here also.

A large number of changes that have been made, but upward compatibility is assured; most lines of VHDL'87 source code will run immediately in the VHDL'92 environment, and the few remaining lines can be easily corrected.

1.4. DEVELOPMENT PHASE

Independent of the debugging tools, which are proprietary tools and therefore outside the scope of the language, some new features in VHDL'92 can improve the design development phase.

The attribute 'PATH_NAME (see chapter 17) returns a string describing "where we are" in source code. Associating this feature with an assert statement, or possibly with the new report statement (see chapter 27), provides a useful aid for tracing the source code.

1.5. DETERMINISM

VHDL'87 is deterministic if we exclude some marginal constructs such as non-associative or non-commutative resolution functions. VHDL'92 clearly encompasses two new sources of non-determinism.

Due to major requests in system modeling, global variables (called "shared variables"; see chapter 4) have been introduced. Misuse of this feature quickly leads to non-deterministic behavior of the design.

The second source of non-determinism is related to files: it is now possible to OPEN and CLOSE a file. This apparently slight enhancement makes it possible to communicate between processes through files. In VHDL'87, it was theoretically possible to achieve the same «functionality», but one had to use files only from within subprograms, and the semantics was not clearly stated. Actually, most implementations simply disallowed it.

Although these notions are among the most important added to the language, it is easy not to use them in order to ensure the determinism of a design.

1.6. DEVICE MODELING

Some device modeling requires checking the consistency of the port values. It is often necessary, in order to avoid "false alarms," to only perform these checks at a steady state, i.e., just before simulation time advances. The notion of postponed process (chapter 3) allows this, both from within the architecture body and from the entity declaration.

When modeling an existing device, it is desirable to use its real name as an identifier. VHDL'92 relaxes the constraints on identifier syntax (chapter 22) and offers an extended character set (chapter 21) for this purpose.

1.7. FORMAL PROOF

Formal proof concerns are very close to those of synthesis. The problems of pattern recognition are about the same in both domains. (Please refer to section 1.12 of this chapter for further information.)

1.8. IMPLEMENTATION PROBLEMS

It may seem amazing to find implementation considerations in a chapter entitled "Designer's Concerns." These preoccupations are usually left to the tool builder, and the designer ignores them. This is an error. Implementation problems represent a cost, and this cost is paid by the end user, the designer.

For example, the notion of overloading, generally well accepted and popular among users, is one of the most expensive implementation problems of VHDL'87. It burdens front-ends (compile-time) and debuggers (run-time). As a result, the price of tools is higher, good tools appear on the market later and performance (compilation time in this case) is lower than it would be if this notion did not exist.

Are there new notions of this kind in VHDL'92? We hope not. Implementation impact was taken into account very early in the language design process and is even one of the guidelines (section 2.4.1.13). Nevertheless, new concepts such as the notion of signature (chapter 23) do not simplify the source code analysis and probably represent a significant cost on implementation.

1.9. PORTABILITY

Although VHDL'87 is a standard, it is not fully portable. For example, real numbers may have different degrees of accuracy depending on implementation, and computation results may be affected. Although it may be difficult to workaround real number portability, there are numerous cases in which non-portability depends on modeling practices and can be avoided.

VHDL'92 clearly states that generics for roots (see section 36.2) are legal in VHDL but that a given implementation may prohibit them. This is a portability issue, and the correct modeling practice is not to use generics on roots.

On the other hand, reading a file during the elaboration phase in VHDL'87 was not portable, due to unclear assertions in the LRM. VHDL'92 explicitly allows for such a modeling practice (see chapter 10) which, in fact, becomes portable.

The concatenation mechanism (&) of VHDL'87 was so restrictive in its use with subtypes that some vendors allowed some digression. The result was a more suitable (but not portable) concatenation operator. In chapter 28, the new strategy of the operator & is detailed. Its new definition makes it both usable and portable. Compatibility with VHDL'87 is another problem.

1.10. READABILITY

VHDL'87 source code was severely criticized for its verbosity. VHDL'92 does not fundamentally differ in this regard but some improvements should be mentioned.

Although the powerful configuration mechanism (component declaration, instantiation, and configuration) has been retained, a notion called "direct instantiation" (see chapter 5) has been introduced for simple cases. This new kind of configuration allows the notion of component to be ignored in some cases, and thereby makes unnecessary the writing of the component declaration and configuration.

The restriction on the reading of output ports in VHDL'87 very often led to the creation of a local variable or a global signal used only for this purpose[5]. This does not enhance the readability of the code. A new attribute DRIVING_VALUE (see chapter 13) allows access to the contributing values of such ports and should eliminate artificial variables and signals.

A new identifier syntax (chapter 22) gives more freedom in their choice and should also contribute to a more readable code.

A new bracketing (chapter 29), in its extended form, should also result in more readable source code, because it is syntactically more consistent.

1.11. SIMULATION EFFICIENCY

A major criticism of VHDL'87 was the lack of a sufficiently predefined environment. VHDL'92 fills this gap by offering new operators (xnor (chapter 12), shift (chapter 11)) and attributes (IMAGE (chapter 16), VALUE (chapter 16), DRIVING_VALUE (chapter 13)) that are "wired" into the simulator, and consequently may be efficient in terms of performance.

The attribute DRIVING_VALUE also contribute to simulation efficiency by preventing the creation of local variables or global signals when reading an output port (chapter 13).

The new "unaffected" clause of the concurrent signal assignment statement (chapter 26) prevents the execution of useless transactions.

The "reject" clause (chapter 18) also avoids signal assignments.

In specific cases (for example, at the system modeling level), global variables (shared variables chapter 4) allow replacement of global signals. The cost of a variable in terms of simulation performance is more acceptable than the cost of a signal (a driver to store, a scheduler to manage).

---

[5] Note that changing the port mode from **out** to **inout** is not a good practice, because it modifies the external view for an internal purpose. This is a poor choice in terms of documentation and readability.

The foreign mechanism (chapter 8), can also be used to call external programs or models, thereby accelerating a small part of code requiring optimization.

The notion of postponed process (chapter 3) allows certain checks to be performed on signal values only once per time point, during the last delta. This avoids often senseless multiple checks each time a signal value varies during certain delta delays.

1.12. SYNTHESIS

We can summarize the concerns in the synthesis domain as follows:

- How can a pattern be recognized?
- How can a constraint be expressed?
- What is predefined?
- What are the modeling practices?

One of the main demands of synthesis users was to add a finite-state machine statement to the language. This demand has not been taken into account in VHDL'92. Indeed, many methods still exist to describe a finite-state machine (e.g.process with case statement, guarded blocks, etc. [BER]), and adding a new one would be redundant. Nevertheless, VHDL'92 provides features for annotating any portion of code (see chapter 7), and it is possible to indicate that such a part of a program is supposed to be an automaton, a RAM, or an ALU. Furthermore, the notion of a group (chapter 7) allows global annotation of a collection of entities (e.g.blocks) in order to specify a piece of information (e.g."they must not be synthetized"). The introduction of the new predefined attribute 'DRIVING_VALUE (see chapter 13) will also simplify recognition by avoiding local variables whose only purpose was to provide a means of reading output port values.

Putting a constraint in the language was already possible in VHDL'87 by using the attribute mechanism. This mechanism has been extended in VHDL'92. For example, due to the notion of signature (see chapter 23), one can attribute an enumerated type subelement in order to specify, for instance, a specific coding. State coding is an application. The generalization of labels (see chapter 35) to all sequential statements also allows the use of attributes for specifying a timing constraint or a resource allocation for this operation. Of course, these practices have to be standardized, but the possibility is now offered by the language.

[BER] VHDL Designer's Reference, Kluwer Academic Publishers 92, Bergé Fonkoua Maginot Rouillard

The predefined environment has been enriched by the addition of new operators like xnor (chapter 12) or shift and rotate operations (chapter 11). No doubt these operators will be overloaded in synthesis environment packages. The semantics will be the same.

Modeling practices are also a sensitive topic in the synthesis field. As is the case for the logic type definition (package STD_LOGIC_1164), they will rely on standard packages that are not defined within the LRM. One of the language design guidelines explicitly disallows the introduction of an application-specific package within the standard (see section 2.4.1.12). A very active standardization group is currently working towards issuing standard packages for synthesis, but this work is outside the scope of the VHDL'92 standardization.

1.13. SYSTEM MODELING

Programming languages are (or were, before VHDL) commonly used for system modeling. VHDL'92 adds certain aspects of programming languages to VHDL.

Global variables (called "shared variables" (chapter 4)) are among these additions.

Impure functions (chapter 10) are also a real help for system designers. By allowing functions to have a memory or to perform read/write operations on files, this feature introduces real non-determinism. Seen as a major problem at a lower level of design (i.e., the gate or RTL level), this property is very useful at the system level, where synchronization problems are not the main issue and where the concerns are to profile a model, to make a stochastic simulation, to find bottlenecks, etc. In this world, determinism is overspecification.

2. SPIRIT OF VHDL'92

2.1. INTRODUCTION

According to the IEEE by-laws, VHDL should to be reballotted every five years. Reballoting provides an opportunity to clean up the language, to improve some features and to add new concepts. User requirements, in quantity as well as in quality, has played a decisive role in justifying these changes.

The Issue Screening and Analysis Committee (ISAC), composed of people involved in VHDL implementations, has for a long time been responsible for the identification and correction of ambiguities and errors in the VHDL Language Reference Manual (LRM).

In a more informal way, several documents and publications ([AUG], [AIR]) have in the past brought up issues about VHDL and discussions about possible changes in the language. For example, the VHDL Annotated Language (VAL) team has identified several drawbacks of VHDL and has worked to improve them.

Also, at the beginning of the restandardization process, special interest groups produced "wish lists."

User requirements can be classified as follows: bugs and repairs (e.g., TEXTIO, which is ill-defined), improvements (e.g., provide a no-change option for conditional signal assignments), and new concepts (introduce analogue modeling). An additional division can be made between requirements regarding the language itself (remove semicolons) and requirements regarding its use (provide a SIN function).

Some of the requirements, mainly the synthesis-related ones, could be answered by the standardization of "standard" packages, such as the STD_LOGIC_1164 package defining a nine-state multi-value logic type and its

---

[AUG] Hardware Design and Simulation in VAL/VHDL,Kluwer Academic Publishers, Augustin Luckham Gennart Huh Stanculescu,

[AIR] VHDL, du langage à la modélisation, Presses Polytechniques Universitaires Romandes, Airiau Bergé Olive Rouillard.

associated resolution and logical functions (which has already been voted on) or the standard packages of Ada. Such packages could be written in pure VHDL and introduced as part of the standard or of sub-standards. An example of a VHDL application that became a standard is WAVES.

Other requirements included changes in the definition of the language itself. The next chapters introduce in a tutorial manner all the language changes that have been accepted, and highlight the main differences these changes imply in the use of VHDL.

2.2. MOTIVATIONS FOR REVISION

2.2.1. Implementations

Some ideas for modifications/corrections came from existing applications: the VHDL'87 standard had been balloted without any VHDL platform available for testing features and concepts. Many examples of the LRM were simply incorrect. Once real VHDL tools were completed, many problems came to the surface. Obvious bugs and imprecisions were taken care of by the ISAC. More fundamental issues were debated publicly or implemented silently (e.g., some constraints on static objects were relaxed).

2.2.2. New Application Domains

VHDL was designed with simulation semantics; a canonical simulator is described in detail in LRM [12]. The consequence is that the semantics has to be bent for synthesis or other domains.

A classical example is the timing semantics: VHDL has descriptive delays, whereas synthesis tools ask for min-max constraints. VHDL has distributed delays (at the assignment level), and synthesis prefers more global pin-to-pin timings.

Such examples can be found for other application domains (e.g.,system level modeling seeks undeterminism, but VHDL is deterministic).

Of course some VHDL tricks could be and have been used to work around this biasing of semantics, but the standardization committees have been under pressure of these user communities of application domains other than simulation.

2.3. THE FORMAL PROCESS

2.3.1. The Standardization Bodies

Standard can be of several kinds. Whether created from scratch by *ad-hoc* technical committees (Ada and VHDL are good examples), declared *de-facto* by the acclamation of the user community (like C), made public by a vendor (GPIB) or imposed by a major vendor (DOS), successful may standards end in an appropriate national standardization body (ANSI, AFNOR, DIN, etc.). The next step is an international standardization body (ISO or IEC).

VHDL was defined by an IEEE committee and has been submitted to ANSI.

2.3.2. The Standardization Committees

The IEEE has Design Automation Technical Committees, beneath which are Design Automation Standard Subcommittees, where the real (technical) work is done. Very often these subcommittees split again into sub-subcommittees to address a particular problem. The final quantum of these sub committees is not computable from the number of participants since most members share their time between many committees and their company work.

2.3.3. The Standardization Process

The process of restandardizing VHDL has led to more than 280 requirements, submitted by users in a formal way. From these, roughly 200 came from the North American chapter, 80 from the European chapter. The Japanese chapter, just starting, has been active mainly in a reviewing capacity. The requirements have been gathered in one document, the Requirement Document for VHDL'92.

In a parallel way, the result of the work of the ISAC, once balloted by the VASG (VHDL Analysis and Standardization Group), has produced the SOVASG, Sense of the VASG. The not-yet-balloted issues have been inserted as requirements in the restandardization process.

As shown in figure 2.1, these two sources of requests have been the inputs of the standardization process. It is noteworthy that the ISAC requirements, which translate the concerns of implementors, mainly deal with unclear points or inconsistencies in the LRM, while user requirements are proposing new possibilities for the language. The main output of the process is the new version of the LRM, the subject of the vote.

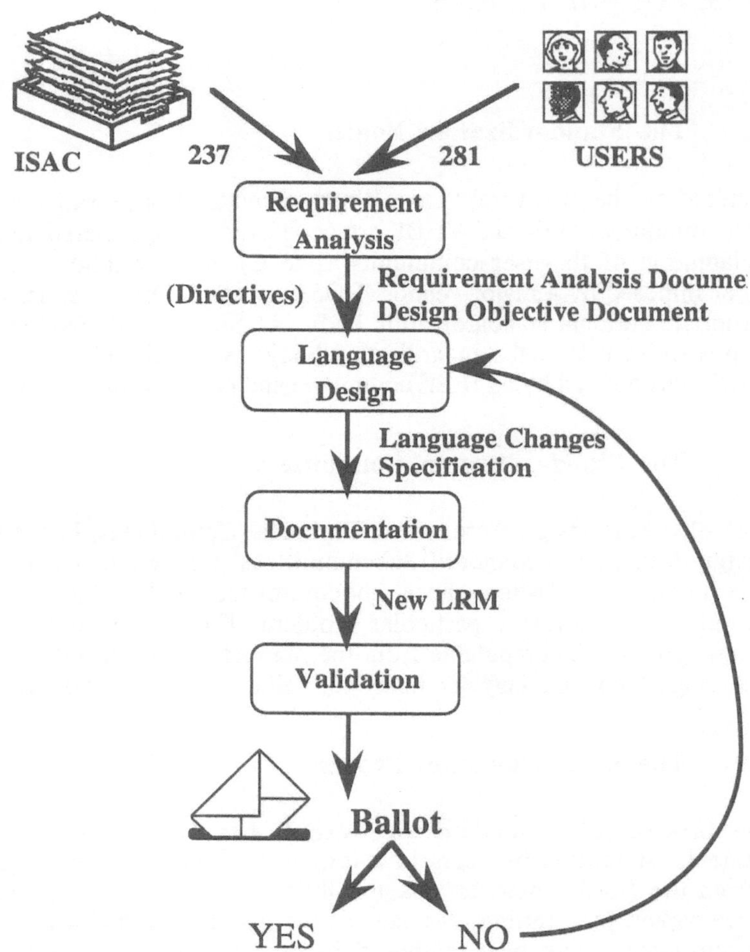

Fig 2.1 Restandardization Process

Although it consists of four main tasks - requirement analysis, language design, documentation and validation - standardization yields other intermediate documents. These will be detailed in a next section.

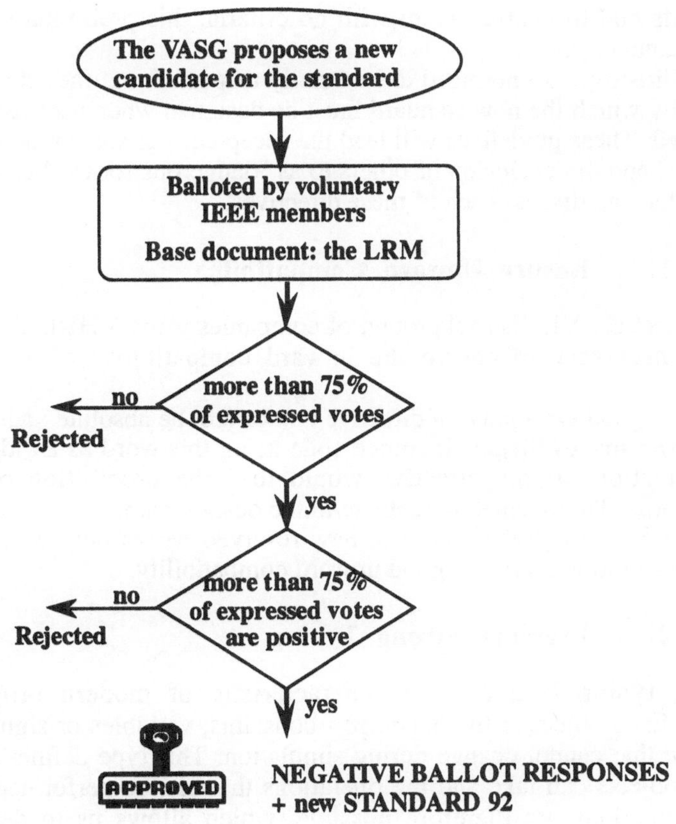

Fig 2.2 IEEE Balloting Process

2.4. THE TECHNICAL PROCESS

2.4.1. Guidelines for Restandardization

One of the main objectives of VHDL'92 standardization was to maintain consistency with VHDL'87 (i.e., not to break the language).

To address this objective, fourteen directives have been established by the Requirement Analysis Committee. Strictly speaking, this task is not part of requirement analysis. Nevertheless, since it seemed difficult to take some 500 requirements and to analyze them with no criteria, this preliminary work has been felt mandatory.

These directives do not need to be strictly respected, but they do define the guidelines by which the new standard must be designed when user requirements are discussed. These guidelines will lead the acceptance of some requirements as "reasonable" and the exclusion of others to as "dangerous for the language."

Let us list and discuss each of these directives.

2.4.1.1. Ensure Upward Compatibility

To protect the VHDL background of companies using VHDL'87, it seemed obviously necessary to ensure the upward compatibility of VHDL'87 to VHDL'92.

Here is a good example of a directive that cannot be absolute : adding a new reserved word makes illegal all source code using this word as an identifier. A strict application of this directive would lead the prohibition of all new reserved words. This seemed unrealistic to the design team.

A good compromise is to create new reserved words only when they are very useful and to guarantee a good upward compatibility.

2.4.1.2. Preserve Strong Typing

Strong typing is a common characteristic of modern programming languages. Each object of the language - constants, variables or signals - has a unique type that cannot change during simulation. This type defines the values that these objects can take and the operations that can be performed on them. Static verifications are therefore possible, which allows us to detect many possible inconsistencies during compilation and so to avoid many run-time errors. Since the cost to correct a run-time error (perhaps by using a debugging tool) is higher than the cost to repair a compilation error, strong typing appears to be a desirable property of a language.

In some cases where types are very close in their usage (but conceptually different), strong typing can be seen as a constraint by some users. Common user requirements request breaking the strong typing between boolean and bit type or between bit vectors and character strings. The present directive clearly tends to discard such requests[6].

Other requests accept the strong typing concept but would like to transform it into a strong typing "à la Pascal." In such a language, the basis of the typing is the structure: two types may be compatible in some cases if they have the same structure. The VHDL notion of typing, similar to that of Ada, is lexical: two types are compatible if they have the same name.

---

[6] However, the bit string literal is now defined in terms of equivalence to a string literal. (see chapter 25)

Although Pascal has been embraced by the users, its typing shows too many exceptions and inconsistencies, especially in I/O primitives, to be a satisfactory example. The Ada typing mechanism seems preferable.

2.4.1.3. Separate Declaration and Functionality

This notion is important in object oriented programming. The declaration usually defines the name of the object as well as the operations that can be performed on it. Separating the definition from the behavior of the object itself (its functionality) provides many benefits. Encapsulation is made easier and extends to good readability of the source code. Readability is improved: the objects that we deal with are well defined, and it is possible to hide some implementation details.

A good example of such a principle is subprogram declaration. A subprogram declaration defines how to call it but nothing about its functionality (except from a documentation point of view, if we give a specific meaning to the subprogram and formal parameter names). This allows us to export subprograms from a package and to hide their realization (body). Deferred constants are another example of this principle.

2.4.1.4. Ensure Unification of Timing Semantics

In VHDL'87, there is only one semantics for time. As is the case with many other features, the corresponding semantics is simulation. VHDL'87 has been designed as a description language more than as a specification one.

An example of such an idea is the **after** clause. This clause enters a time value, associated with a potential future value, into a driver of the target signal. The time value defines the exact date when the current value of the signal will be updated (if nothing else contradicts this). If this value is 27 ns, it does not mean 26.9 or 27.1 ns. An example of mixing two semantics could be to read "after exactly 27 ns" for a simulation purpose and, for example, "after less than 27 ns" in a synthesis context. This is not the case in VHDL'87. The timing semantics is unique (and turned toward simulation).

Other policies might lead to ambiguities in source descriptions.

2.4.1.5. Preserve Determinism

Since VHDL'87 is not deterministic in some exotic cases, it may seem difficult to preserve determinism. In fact, the main sources of non-determinism in VHDL'87 are well known and very limited: resolution function and IOs.

Resolution functions can lead to non-determinism if they are not associative or commutative, i.e., the same set of driving values can lead to different effective (resolved) values. Indeed, such functions have a result dependent on the order of the values in the input array, and this order is itself not guaranteed.

Many constraints are placed on files in VHDL to prevent their use as communication channels between processes. If two different VHDL files correspond to the same logical file, read/write operations without simulator synchronization can lead to nondeterminism.

Even if VHDL'87 is strictly speaking, non-deterministic, it is easy for the designer to know when he or she is using features that can lead to such behavior. The same spirit must be applied to VHDL'92.

Two strongly requested requirements are also potential new sources of non-determinism in VHDL: global variables and analogue related requests. Global (shared) variables allow uncontrolled communication between processes. In the analogue domain, it seems unreasonable to expect a definition of the internal algorithm of the analogue simulator (which is the case for the digital one): thus, differences of simulation results between platforms are likely.

2.4.1.6. Preserve Generality

VHDL is a hardware description language that does not even have a predefined **and** gate (**and** is an operator). Since the early versions of VHDL, the language designers have always chosen to give priority to generality over specificity.

Of course, VHDL'87 is a general language. Its variety of datatypes and levels of description give it a large range of application. This is the result of research on generality during its design.

It is sometimes easy to create the specific notion or the exact mechanism to address a given problem. Further thought can lead to a generalization that can be applied to other cases. The result is a more powerful but perhaps more complex language.

In fact, the present directive often leads to decoupling the language from the hardware notions. For example, the smart notion of the resolution function is a general answer to a request for a mechanism to handle bus conflict arbitration. Starting from a hardware-based need, the answer is a general software solution. And resolution functions may be useful in many more cases than bus modeling.

Generality changes the level of abstraction, and this can be disturbing for many designers. The requests are pragmatic, but the answers are generic. Here is an example:

"Please," say the designers, "give us a mechanism to support pin-to-pin timing."

"OK," answer the language designers, "you will be able to declare an annotation relationship (notion of a group, in chapter 7) between objects and to give them a value. So, it will be possible to declare a relationship between ports, to give it a value of type TIME, and to use it for pin-to-pin timing."

The language is more powerful, but the expertise needed to use it increases also. Generality is a good directive but does not simplify the use of the language.

2.4.1.7. Preserve (at Least) a Gate (Circuit) System Scope

VHDL'87 covers the range from system level to gate level. It is obvious that the new version can extend the scope but should not reduce it.

2.4.1.8. Preserve Intermixed Levels of Abstraction

The possibility of mixing different levels of abstraction in the same model is a very powerful characteristic of the language that allows top-down refinements.

It is possible to refine only part of a description (to use a lower level of abstraction) without affecting the consistency of the modeling. For example, a refining step in a common top-down methodology may consist of replacing part of a behavioral architecture by a structural one. This kind of option must remain open.

2.4.1.9. Preserve Concurrency

Preserving a real concurrency means keeping the semantics of a description independent of the order in which the concurrent statements are written. The synchronization mechanism (drivers, delta delays, wait statements) defined in the VHDL'87 LRM guarantees a full and exact concurrency.

VHDL'92 must provide the same (or a similar, upward-compatible) mechanism.

2.4.1.10. Preserve (Improve?) Language Consistency

VHDL'87 is a rather consistent language. Many mechanisms are general and lead to few exceptions. Exceptions can be found, however, particularly in the syntactic domain.

The construct bracketing syntax is not uniform, as shown by the following example:

```
component COMP ... end component; -- no "is" reserved word

entity ENT is ... end ENT; -- and not "end entity;"
```

In the same spirit, some default characteristics of constructs cannot be made explicit in VHDL'87. One example is the inertial mode for signal assignment.

This mode, dual to **transport** mode, is the default mode of signal assignments but cannot be explicitly stated.

The VHDL'92 language designers took care not to add very specific rules or mechanisms and, when possible with regards to upward compatibility, also tried to correct some inconsistencies.

2.4.1.11. Preserve (Improve?) Portability

Portability has been a matter of discussion in VHDL'87. The first experiments quickly showed that writing portable VHDL source code was not an obvious process.

Some types such as INTEGER, REAL, and even TIME, were not portable. Some functionalities, such as textual input/output operations with TEXTIO, also led to problems. Source codes targeted to synthesis are also less portable: they can be seen as a proprietary language based on user-defined attributes written between VHDL source code lines.

If we consider the mobility of the CAD market, portability is a very acute problem. Users want to preserve their investment in source code.

Preserving the VHDL'87 code was the most fundamental wish of the VHDL'92 language design team.

2.4.1.12. No Application-Specific Packages in the LRM

The wording of this directive has always been ambiguous. Of course, some application-specific packages must be defined and recognized as standardized packages. The meaning of this directive is to prohibit such a package from appearing in the LRM.

Application specific packages have too specific use to be part of the VHDL language. The language is complex enough to extract from it whatever is not used by a majority.

Another reason is more pragmatic. The timing of restandardization was quite short, and it was essential to focus on the language and not on the applications. Moreover, VHDL mechanisms such as packages allow very simple decoupling of these concerns from the language. WAVES is an excellent example of such a decoupling.

Some packages are being standardized by the synthesis, analogue and formal verification working groups.

2.4.1.13. Minimize Implementation Impact

Tool builders have made a great investment in VHDL tools. Due to the complexity of VHDL, many tools are very new. A consistent strategy in VHDL'92 language design has been to give priority to the solutions that minimize the implementation impact.

2.4.1.14. Maximize Implementation Efficiency

Implementation efficiency is always a sensitive problem. To compete with VHDL, other HDLs often plead for efficiency. VHDL defenders say that it is difficult for them to discuss VHDL generality or consistency. Nevertheless, such an argument is the central preoccupation of certain designers and must not be neglected.

New concepts or features that can help to optimize implementation efficiency must be studied. A good example of such a concept is global (shared) variables. At a very high level of description, system designers would prefer to use variables instead of signals. The cost of signals for the implementation is large (drivers, resolution functions) and is unnecessary at this level; the description is so far from hardware that only global variables such as these available in conventional programming language are needed. This directive was the main asset of language designers who wanted to introduce global variables in VHDL.

2.4.2. Requirement Analysis

The analysis of the requirements has shown a great heterogeneity in the abstraction level, as well as in the application domains. The first work of the Requirement Analysis team was to bring the whole set of requirements to the same level of abstraction.

This work was undertaken during some working meetings and through electronic mail. During this phase, some requirements were not clear to the working group. These fell into two categories:

- Requirements without enough information, without the requirement itself, with few or no comments, or without an identified author, many of which came from a brainstorming meeting of a special interest group. These have not been studied further.
- Requirements falling into a very specific application domain, in which the analysis committee had no expertise. In this case, the submitter was requested to provide more information, or, when possible, to champion his requirement. This occured primarily with requirements related to formal proof and synthesis.

Finally, a classification of design objectives was made by the Requirement Analysis Committee, based on the technical contents of the requirements and their respect of the guidelines. The classes are Design Requirement, Design Goal, Study Topic, and Deferred.

No consensus could be reached on the study topics, and a negative consensus was found on the discontinued objectives. Nevertheless, it happened that somebody volunteered to champion a study topic. Some of these were promoted to design goals or design requirements as the result of this action.

The design requirements were the primary concern of the Language Design Team. The design goals were also considered important but, due to time and manpower consideration, might be not implemented.

The set of classified design objectives can be found in the Design Objective Document.

2.4.3. Language Design

The degree of freedom of this design phase, speaking in terms of choices, concerned the design goals.

The frontier between design requirement and design goal was quite fuzzy, and the decision to implement or not was based on time and available manpower. Even the frontier between some design requirements and pure miracles was not so clear.

The Language Design phase should have ideally implemented all the design requirements. However, some technical problems were identified only at this point. In some cases, getting around some issues led to such complexity in the language that the design team finally gave up.

Moreover, the language design for one particular requirement completely missed the intent of the submitter, and the objective had to be partially retargeted.

2.4.4. Documentation

2.4.4.1. First Considerations

Many people expect VHDL'92 to introduce a new concept for documentation. The VHDL'87 Language Reference Manual did not satisfy users, and they found it unreadable. Some of the misunderstandings arose from two questions: what is the LRM supposed to do, and who is supposed to read it?

The LRM is *not* a tutorial. It is a contract between the user (who is supposed to be represented by the IEEE Society in this case) and each tool builder who wants to match the standard. As a contract, it must be unambiguous rather than convenient to read, exactly like an insurance policy.

Of course, clarity does not necessarily mean ambiguity, but many niceties in style can lead to controversy. For example, when a concept is complex or related to different parts of the document, some duplications of the definition (sometimes with different approaches) appear to be reader friendly. On the other hand, such repetitions can lead to inconsistencies: different phrasings may create questions such as "Why? Where is the subtlety?"

As for VHDL'87, some translations of this manual (into Japanese, German, French or Spanish) will probably be allowed by the IEEE. Nevertheless, the only contract remains the English version, and this fact is stated on all other versions.

To sum up, what can the designer expect from the Language Reference Manual? If he or she is a beginner, nothing! Comprehensive courses are available and must be the prelude to any attempt to read this manual. Any other option, even for skilled engineers, will be risky.

As an advanced user, the designer may find in the LRM unambiguous answers to specific questions about the language. Of course, the VHDL tool-builder will necessarily use this manual in order to develop any analyzer and/or simulator satisfying the standard.

2.4.4.2. Documentation Process

The documentation phase consisted mainly of writing the LRM. Since VHDL'92 is a second release of VHDL, some incremental information has been made available, mainly during the balloting phase, such as identification of the changes in the language.

The new LRM is self-sufficient, which means that the "old" LRM has become obsolete.

The new LRM includes an appendix identifying portability issues.

2.4.5. Validation

The validation phase had quite ambitious goals: to validate the LRM against itself (consistency) and against user's requirements and ISAC issue reports. Some general rules were defined as properties that must be preserved by the updates made to the VHDL'87 language. The validation team had the task of tracking down all the changes that could violate any of the general rules. Fourteen general rules were defined, corresponding to the refinements to the guidelines presented in section 2.4.1. Unfortunately, these properties could not be proven using automated means, because no formal semantics is defined for VHDL and also because today's automatic theorem provers still apply only to problems of limited size. So, the VHDL'92 validation process included a combination of automatic and manual checking. It comprised three main components [ZAM]:

Formalizing: it covered specific aspects of the language such as library management, key elaboration features, and the modifications affecting the

[ZAM] Validation emails - available from A.Zamfirescu, chair of the task.

simulation kernel. Different mathematical formalisms were introduced to model these features. For example, one of these formalisms was used to prove that the postponed processes added to the language were well defined.

Semi-automatic inspection: this included the definition of a glossary and the use of a relational database to store the links to the parts of the LRM that might be affected by specific language changes. These parts had to be manually checked for consistency and for compliance with the general rules. Furthermore, a test set of VHDL examples was developed to make sure the language changes matched the design requirements.

Feature prototyping: this included the rapid development of prototypes to prove the feasibility and the consistency of the new features of VHDL'92. In particular the problems of portability were checked by considering results achieved by different valid prototype implementations of the same features.

All the problems discovered by the validation team were submitted to the language design team for new modifications of the language.

2.4.6. Negative Ballot Responses

The IEEE bylaws request that an appropriate response be given to any negative vote of an IEEE member. The VASG has even decided to extend this request to all comments.

The Negative Ballot Responses is an interesting document: people voting "No" have to explain why, and defenders have to elaborate on this comment. The result is valuable for evaluators and technical managers.

Introduction

=> ***New Simulation Mechanisms***

New Structuring Mechanisms

New Interfacing Mechanisms

New Predefined Operators, Functions & Attributes

Slight Enhancements

Language Simplifications

Clarifications

Annex

3. LAST-DELTA ACTIVATION: POSTPONED PROCESS

LRM REFERENCES: 9.2

3.1. BACKGROUND

The concept of delta delay - simulation step seen as an infinitesimal delay - is the VHDL artifice to enforce causality in simulation. A time point of simulation consists in a variable number of delta delays that are necessary to the simulation semantics. It may happen that a designer wants to ignore what occurs during these delta delays in order to focus only on the "steady state" at the end of the time point, i.e., during the last delta.

Figure 3.1 gives an example of an assertion statement that generates "false alarms" each time the value of signal A changes.

3.2. DESCRIPTION

VHDL'92 introduces the notion of "postponed" process. A postponed process is only activated during the last delta of a time point.

The example given above can be easily solved by using a postponed assertion statement, which is also a process. Such a statement will be performed once per time point, during the last delta. When this occurs, the expression

A=B is evaluated, but the steady state has been reached, and the assertion is performed if necessary. The postponed process mechanism is described in detail below.

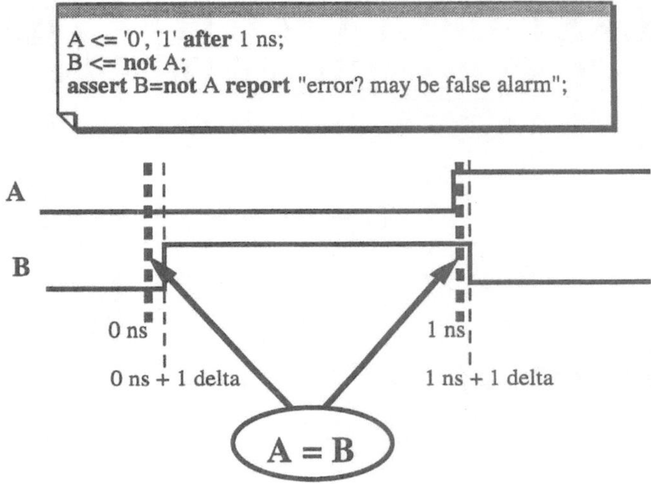

Fig 3.1 False Alarm Generation

In the general case, a process is sensitive to a set of signals, possibly associated with conditions (**wait on** ... **until** ...;). Each time an event occurs on one of these signals (if the condition optionally associated is true), the process is immediately activated. If this process is a postponed process, activation is deferred to the last delta delay of the time point (i.e., just before the simulation time advances). As shown later in section 3.3.2.1, the evaluation of the optional condition will not be redone during the last delta activation. If more than one postponed process have to be activated during the last delta, the order of these executions is implementation dependent (i.e., the LRM does not define any order) but not really relevant for the designer.

There are two main restrictions to the use of such a process:

- Since it is activated during the last delta, a postponed process is not allowed to create new deltas during the same time point. The only way for a process to generate new delta delays is to perform a zero-delay signal assignment (one waveform element equal to zero) or to wait for a null time-out (0 ns). These two possibilities are prohibited and will be detected by the compiler or at run-time.

  It is important to note that a postponed process is not restricted to being a passive process; any non zero-delay assignment is possible within it.

- If such a process is sensitive to more than one signal, there is no way to know which event caused the process to be activated during the last delta. The value of the predefined attributes on signals cannot be used, not because it is an error to use them, but because they are simply meaningless (except if they appear in the **until** clause expression, which is evaluated during "normal" delta delays).

The three following concurrent statements are mapped into equivalent postponed processes if preceded by the reserved word **postponed** (just after the optional label):
- assertion statement
- concurrent procedure call statement
- concurrent signal assignment (conditional or selected forms)

Note: It is possible for documentation reasons to repeat the reserved word **postponed** at the end of a *postponed* process. The new process syntax is as follows:

```
[process_label : ] [ postponed ] process [ ( sensitivity_list ) ] [ is ]
        process_declarative_part
begin
        process_statement_part
end [ postponed ] process [ process_label ] ;
```

3.3. USING THE FEATURE

3.3.1. Examples

3.3.1.1. Assertion Statement

The assertion statement is the most obvious application of the postponed process mechanism: it performs the checks only when the "steady state" of a given time point has been reached. The first example of this chapter can be easily solved using the following line:

```
postponed assert B=not A report "real problem at the steady state";
```

Not performing multiple checks during the same time point not only avoids false alarms but also saves simulation time (less process activations and less expression evaluations).

3.3.1.2. Process Statement

Postponed process can be very useful to get rid of false alarms in timing checks. For example, let us consider the pulse-width check on a clock signal. The problem is to check that the length of time between two changes of state is always comprised within given bounds.

The change here is considered in terms of the steady state. We want to consider only the ultimate final value of the clock signal in a given cycle. In this regard, we will say that a steady change or event has occurred on a signal if its value as evaluated during the last delta is different from its value as evaluated during the last delta of the previous simulation time point. To detect such an event, the following subprogram, when executed by a postponed process or procedure statement, is sufficient:

```
procedure WAIT_STEADY_EVENT(signal S : BIT) is
        variable LAST_VALUE : BIT :=S;
begin
        loop
            wait on S;
            exit when S/=LAST_VALUE;
        end loop;
end WAIT_STEADY_EVENT;
```

Notice that we store the previous value of signal S before waiting for an event. This is because we cannot be sure after a wait statement in a postponed process that the signal that caused the process to resume is still active. The attribute 'EVENT cannot provide this information because its value does not extend to more than one delta cycle.

Coming back to the pulse-width check, we can state it as follows:

```
        constant MIN_CYCLE_TIME : TIME := 50 ns;
        constant MAX_CYCLE_TIME : TIME := 80 ns;
LBL : postponed process
        variable LAST_CHANGE_DATE, PULSE_WIDTH : TIME;
begin
        WAIT_STEADY_EVENT(S); LAST_CHANGE_DATE :=NOW;
        loop
            WAIT_STEADY_EVENT(S); PULSE_WIDTH :=NOW-LAST_CHANGE_DATE;
            LAST_CHANGE_DATE :=NOW;
            assert (  (PULSE_WIDTH>=MIN_CYCLE_WIDTH) and
                      (PULSE_WIDTH<=MAX_CYCLE_WIDTH) )
            report "Pulse check error" severity ERROR;
        end loop;
end postponed process;
```

3.3.1.3. Concurrent Procedure Call Statement

Timing checks are very frequently encountered in modeling applications, and it may be desirable to encapsulate them in procedures so that they can be reused easily. Furthermore, in VHDL'92, a procedure call can be **postponed**. The previous process can be replaced by the postponed procedure call:

```
LBL : postponed CHECK_PULSE_WIDTH(  S,
                                    MIN_CYCLE_WIDTH,
                                    MAX_CYCLE_WIDTH);
```

where CHECK_PULSE_WIDTH is defined as

```
procedure CHECK_PULSE_WIDTH(     signal S : BIT;
                                 MIN_CYCLE_WIDTH, MAX_CYCLE_WIDTH : TIME) is
    variable LAST_CHANGE_DATE, PULSE_WIDTH : TIME;
begin
    WAIT_STEADY_EVENT(S);
    LAST_CHANGE_DATE := NOW;
    loop
        WAIT_STEADY_EVENT(S);
        PULSE_WIDTH :=NOW-LAST_CHANGE_DATE;
        LAST_CHANGE_DATE := NOW;
        assert ((PULSE_WIDTH >= MIN_CYCLE_WIDTH) and
                                     (PULSE_WIDTH <= MAX_CYCLE_WIDTH))
                                     report "Pulse check error" severity ERROR;
    end loop;
end;
```

3.3.1.4. Concurrent Signal Assignment

A postponed concurrent signal assignment statement should not assign a value with a 0 ns delay. These statements can be used to model clocks, especially when the cycle time is provided by a generic parameter. In that case, the postponed signal assignment will produce an error during the initialization phase if the relevant delay evaluates to 0 ns. Of course, a check could be done using a simple assert statement!

```
entity CLOCK is
    generic   (CYCLE_TIME : TIME);
    port      (CLK : inout BIT);
end;

architecture DATA_FLOW of CLOCK is
begin
    postponed CLK <= NOT CLK after CYCLE_TIME/2;
end DATA_FLOW;
```

3.3.2. Some Traps

3.3.2.1. My Debugger Is Crazy

```
postponed process
begin
        wait on S until COND1;
        ...
end process;
```

When debugging the previous postponed process, during the last delta of the time point, you may be surprised to discover that the condition COND1 is false. Do not send your simulator back to its vendor; this situation is perfectly possible. When your process has been activated at least once during some simulation step (i.e., delta delay) of the current time point, the condition *was* true. This does not mean that this condition is still true during the last delta.

In the same way, do not have high expectations that S'EVENT will be true during the same debugging phase.

3.3.2.2. When Will It Wake Up Again?

```
postponed process
begin
        wait for 10 ns;
        ...
end process;
```

This case has been clearly described in the LRM. This process will wake up in 10 ns during the last delta of the corresponding time point.

4. SHARED (GLOBAL) VARIABLES

LRM REFERENCES: 4.3.1.3

4.1. BACKGROUND

Since the term "global variable" has led to intense debate throughout the standardization process, the more neutral term "shared variable" has finally been chosen. The concept itself has not changed: a shared variable is a variable that can be accessed (read or written) by more than one process.

The VHDL'87 determinism is based primarily on the notion of a signal as the only way to communicate information between processes. For a designer at the RTL or gate levels, a shared variable can appear to be an unnecessary way to break this determinism. The need for global variables is indeed real, but it is more related to system-level modeling, and can be split into two categories of problems:

• Object-Oriented Programming is a good way to structure an application. Packages are very popular in such a programming and can be seen as objects. According to a now classical terminology, such an object may have a state (its internal state) that can be changed by the use of subprograms. This state is the memory of the package; it is remanent data.

In modern programming languages, this state is stored in the package body (or its equivalent) as a remanent variable, which prevents any attempt to modify it by other means than the subprograms of the package. If a need arises to know this state from outside the package, then a specific subprogram must be written to read it.

In VHDL'87, global variables in a package are not allowed but global signal are. Why is this not sufficient?

A good example of OOP is the representation of a stack as a package. Three subprograms PUSH, POP, and TOP can be exported by the package, and we might think that a global signal STATE could represent the contents of the stack. Oops! The semantics of a signal makes this usage very hazardous; for example, two calls to PUSH within the same process, without wait statement in between, will have the same effect as only one call (the second one). The driver-editing mechanism will erase the effects of the first one.

It is not even possible to schedule a wait statement after each call to one of the primitives, because:

- a "**wait for** 0 ns;" statement may be undesirable. In particular, it changes the driver editing for other drivers; and
- it does not solve the problem of concurrent access from several processes: the state might be accessed from many processes at the same delta time. If we PUSH two times, we want two values to be pushed, rather than having a resolution function performed to resolve the supposed "conflict."

Actually, the whole problem comes from the fact that a signal has only one effective value at one given simulation cycle (i.e., during the same delta delay). A variable, in contrast, may have many different readable values in sequence, at the same time point.

- At the system level, designers were using conventional programming languages before VHDL. For their modeling, they commonly use global (or shared, static) variables. At this level, determinism is not a quality: the question is to launch random input patterns, to profile the design and to collect data for statistics. The use of global signals instead of global variables simply does not work.
- The use of signals where only variables are needed is bad for performance: the implementation cost of a signal (driver, current value) is much higher than that of a variable. As seen in the example of the STACK package, the current value of a signal is not immediately updated when the signal assignment is performed. It costs at least one delta delay to do so. The systematic use of artificial synchronisation points (typically "**wait for** 0 ns;" statements) also contributes to bad performance, heavy writing, and, worst of all, a design burdened with nasty workarounds.
- It is often the in-house modeling practice to associate hardware meanings with signals.

To solve these kinds of problems, *shared* variables have been introduced in VHDL'92.

4.2. DESCRIPTION

A shared variable can be declared by using a classical variable declaration preceded by the reserved word **shared** in any declaration region except subprograms and process declarative parts (where "normal" variable declarations are possible).

```
shared variable MY_VARIABLE : MY_TYPE := MY_INIT_VALUE;
```

Using the conventional visibility rules, this variable can be accessed (read or written as a "normal" variable) from any region within any processes or procedures.[7] This notion will be presented below.

4.2.1. Determinism Issues

Two different levels of non-determinism can be introduced by acceding without any restriction to shared variables:

• A first level is related to the model itself: the updating of a variable becomes dependent on the order of execution of processes; this order may neither be specified nor may it be guaranteed the same for different runs of the same model.

• The second level is more subtle. VHDL has some good properties allowing for efficient implementations on massively parallel computers.

In such a case, shared variables will appear as shared resources for two or more processors. If two processors write the same variable at the same time without protection (for example, a record type with multiple fields), the result will be not only unpredictable - which, after all, may be seen as the goal of this feature - but also inconsistent.

Let us assume that a number is stored using two bytes, each being accessible by an atomic operation. Let us also assume that process ONE is willing to increment the number, and process TWO is willing to decrement it.

It may happen that the execution of operation, for some reason, be this one:

---

[7] Except if that procedure is called within a non-impure function. See impure function concept in chapter 10.

The Number is 01FF
Process ONE increments it (low byte first) 0100
Process TWO decrements it (the whole) 00FF
Process ONE strikes back (high order byte) 02FF

After an increment and a decrement of 01FF, it would be acceptable, in terms of data integrity- to find 1FF, or even 200 or 1FE (we may accept to pay the price of non-determinism and concurrency by ignoring which operations have actually been done at a given time). 2FF is obviously unreasonable.

We are facing the classical problem of maintaining data-integrity in operating systems. This "major" non-determinism *has* to be taken care by the designer and the implementor.

4.3. USING THE FEATURE

4.3.1. Open Traps

Shared variables have led to long-term discussions among VHDL'92 designers. They are, along with (non-commutative or non-associative) resolution functions and (closed and reopened) files, another but easier way for non-determinism to enter VHDL.

A simple example can illustrate this point. If we declare a shared variable in an architecture body, this variable can be accessed by each and every process within this architecture. The update of its value, as a classical variable assignment, does not refer to a synchronization point (**wait** statement). This leads to the possibility of strange behavior.

```
architecture NON_DETERMINIST of EXAMPLE is
        shared variable COUNT : NATURAL;
begin
        P1 : process
        begin
          COUNT := 1;
          wait;
        end process;

        P2 : process
        begin
         COUNT := 2;
         wait;
        end process;
end NON_DETERMINIST;
```

At the end of the simulation, nobody can say if the value of COUNT is 1 or 2. This value depends on which process wrote the last value in COUNT. Moreover, this value can be different from one implementation to another, and even for two different simulations within the same implementation.

A good practice is to use shared variables for the purpose for which they were introduced (in system design, and for handling states in Object-Oriented Programming design), and to quickly forget them in other cases.

4.3.2. Reasonable Uses of Shared Variables

4.3.2.1. OOP Modeling

If we try to write *in-extenso* the STACK package cited before, we could end up with this kind of solution:

```
package STACK_OF_INTEGER is
        procedure PUSH (WHAT : in INTEGER);
        procedure POP;
        impure function TOP return INTEGER;
end STACK_OF_INTEGER;

package body STACK_OF_INTEGER is

        constant MAXSTACK : POSITIVE := 1000;
        type STACK_TYPE is array (0 to MAXSTACK) of INTEGER;
        shared variable STACK : STACK_TYPE;
        shared variable INDEX : POSITIVE := 0;    -- the first free position

        procedure PUSH (WHAT : in INTEGER) is
        begin
         STACK(INDEX) := WHAT;
         INDEX := INDEX+1;
        end PUSH;

        procedure POP is
        begin
         INDEX := INDEX-1;
        end POP;

        impure function TOP return INTEGER is
        begin
         return STACK(INDEX-1);
        end TOP;

end STACK_OF_INTEGER;
```

It must be noted that no error-recovery mechanism is implemented in this example. A real package would have to take care of all strange situations (getting the TOP of an empty STACK, overflows and underflows, etc.) by introducing assert statements or by returning a status.

Furthermore, the software oriented reader will notice that a feature is missing: i.e., the possibility of making this kind of package generic with reference to the type contained in the stack. We will have to define a STACK_OF_REAL or a STACK_OF_BIT if we need them: such packages will be pure copies of this one, with a textual replacement of INTEGER by REAL or BIT. If we are unlucky enough to need two stacks of INTEGER, then we need to duplicate the packages with another name.

4.3.2.2. Conventional Programming Techniques

A rare but remarkable use of global variables is to model the temperature of a design. In many places, the "after" clause may depend upon some function of the temperature.

```
S <= A and B after DelayForAnd(Temperature);
```

This temperature, up to now, could only be a global constant or a global signal.

Using a signal has three drawbacks:

• First, it brings with it the propagation semantics of all signals, which means that a change of temperature would be submitted to driver editing and to resolution.

• Second, it brings the determinism of signals: making the behavior dependent on the temperature, which itself depends on statistics derived from the behavior, is not a deterministic simulation. Rather it has to do with system-level modeling and monitoring.

• Third, it is often recommended to dedicate signals to hardware semantics. Since we do not want to synthesize the temperature, the only solution left is to use global constants.

Using a constant has also a drawback: it is constant! It might indeed be interesting to model the behavior of a circuit whose temperature is a function of its own activity (for example, a RAM).

In VHDL'92, the temperature may be stored in a shared variable and kept rather safe (although non-deterministic) if updated and read with the semantics of a temperature. In particular, no value for signals should be computed from the temperature, only delays.

Introduction
New Simulation Mechanisms
=> ***New Structuring Mechanisms***
New Interfacing Mechanisms
New Predefined Operators, Functions & Attributes
Slight Enhancements
Language Simplifications
Clarifications
Annex

5. DIRECT INSTANTIATION

LRM REFERENCES: 9.6.2

5.1. BACKGROUND

Until now, the only way to "use" an entity within a design has been to declare a component, to instantiate this component, and to configure it with the given entity. Figure 5.1 illustrates these operations. The notion of a component is one of the most complicated features to be understood by the VHDL beginner. Even if this mechanism is very powerful in the general case, it can appear as very verbose in simple cases and does not improve the readability of the whole design (for designers as well as for dedicated tools). Furthermore, other HDL vendors that do not propose so sophisticated a mechanism also solve many problems and may easily claim that VHDL is too verbose.

5.2. DESCRIPTION

It is always possible to use the component notion in VHDL'92 as well as the configuration notion, but VHDL'92 allows for an optional simplification called "direct instantiation."

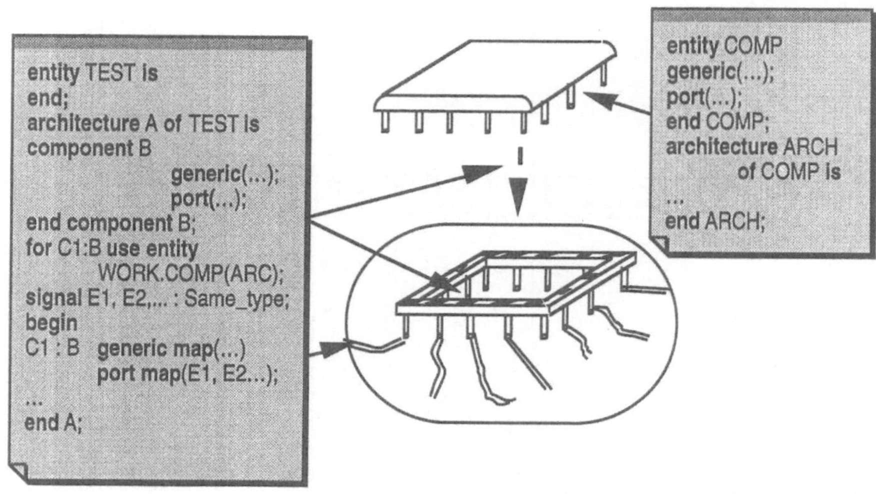

Fig 5.1 Component Declaration, Configuration, and Instantiation

When performing a direct instantiation, the component instantiation statement consists of an entity/architecture pair (or a configuration unit name; this will be illustrated later) instead of a component name, preceded by the reserved word **entity**. This pair is implicitly configured. The following statement instantiates the entity INVERTER associated with the architecture DATAFLOW as the component instance INV1:

```
INV1 : entity WORK.INVERTER(DATAFLOW) port map (INPUT, OUTPUT);
```

Since no component declaration or configuration is necessary anymore, the number of lines of source code dramatically decreases. Figure 5.2 shows this simplification for the same example as in figure 5.1.

The general syntax of the instantiation of a component is now

*instantiation*_label : instantiated_unit [generic_map_aspect] [port_map_aspect] ;

where instantiated_unit means one of the three following forms:

- The component name, optionally preceded by the reserved word **component** for documentation purposes:

  [**component**] *component*_name

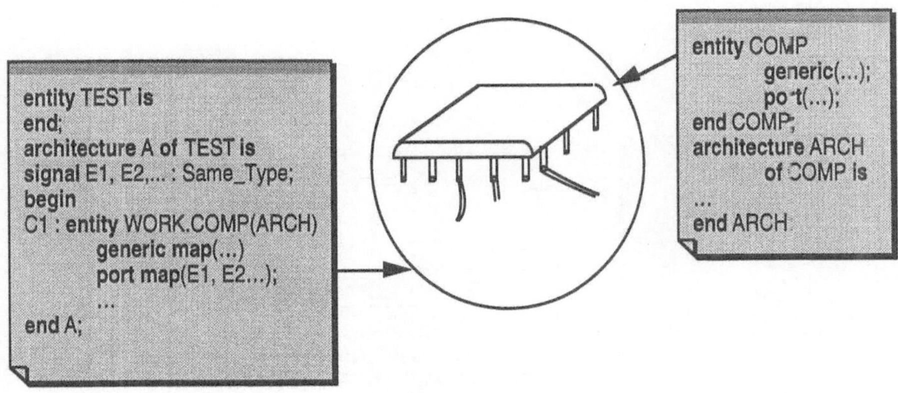

Fig 5.2 Direct Entity Instantiation Mechanism

- An entity/architecture pair that can be optionally reduced to the name of the entity only[8] (see LRM [5.2.2], configuration default binding). This form implies the use of the reserved word **entity**:

  entity *entity*_name [(*architecture*_identifier)]

- A configuration name preceded by the reserved word **configuration**:

  configuration *configuration*_name

5.3. USING THE FEATURE

A designer using direct instantiation takes a risk - namely, forgetting the possibilities of the notion of a component and the power of the configuration mechanism. These two features are essential in term of top-down methodology, generality and reusability of a design. Since direct instantiation is very simple to use, the real question is not how to use it, but when.

5.3.1. Right Use of Direct Instantiation?

• When handling a single instance of an entity/architecture pair, always within the same architecture, and for a single component without any purpose of reusability, direct instantiation can be used without restriction.

[8] In this case, the most recently compiled architecture of the specified entity is used.

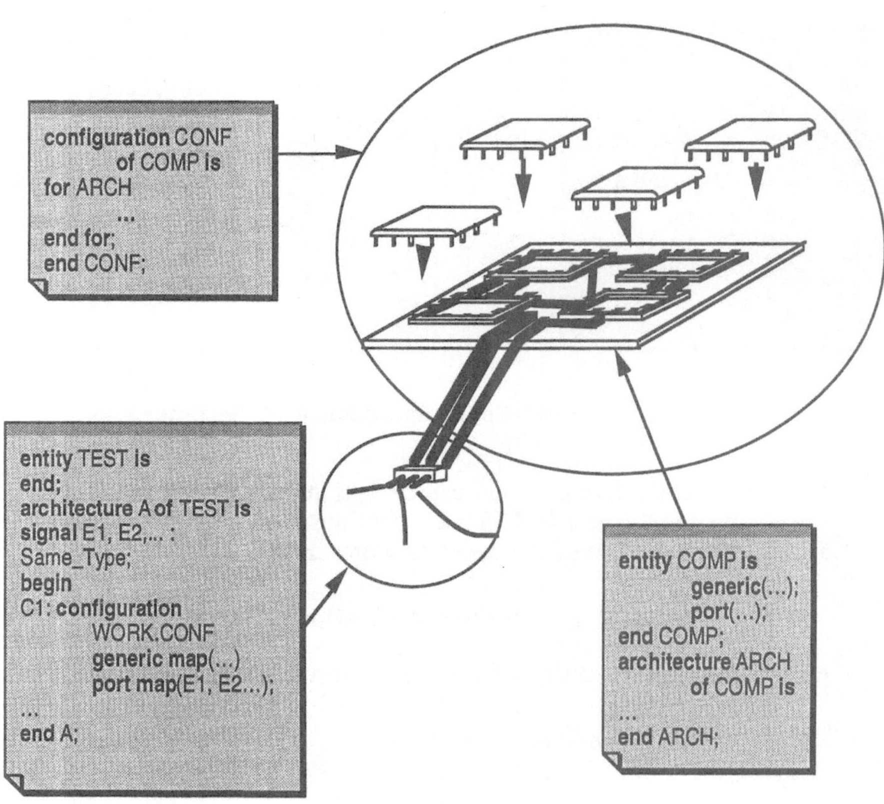

Fig 5.3 Direct Configuration Instantiation Mechanism

As shown in figure 5.2, this kind of situation can often be encountered when defining the test environment of the system. In that case, the system has been designed as an entity/architecture pair, and this entity has ports and generics. The goal now is to embed it in its test environment. This consists in creating a netlist with an instantiation of the system and to generate stimuli, as well as optionally to check the outputs.

• In the main architecture of a system, one can also have a very small number of single "components" (a RAM, a ROM, a CPU, etc.). The use of direct instantiation is also possible and increases the readability of the source code:

```
RAM : entity RAM124(ARCH26) port map...
ROM : entity ROM312(ARCH8) port map...
CPU : entity CPU32(ARCH) port map...
```

Nevertheless, if the architecture contains two or more instantiations of the same "component" (two RAMs), the direct instantiation mechanism will enforce a duplication of the information of the corresponding entity/architecture pair, which is not a good practice.

```
RAM1 : entity RAM124(ARCH26) port map ...
RAM2 : entity RAM124(ARCH26) port map ...
ROM : entity ROM312(ARCH8) port map...
CPU : entity CPU32(ARCH) port map...
```

5.3.2. Wrong Use of Direct Instantiation?

It may seem very appealing to use direct instantiation all the time to avoid the "verbosity" of the component declaration and component configuration specification. But this would not be very wise because of the reusability overhead. This point is illustrated by the example of a simple adder.

Assume that the following units are available in the library NMOS:

```
entity NMOS_AND is
        port( INPUT1,INPUT2 : BIT; OUTPUT : out BIT);
end entity NMOS_AND;

entity NMOS_OR is
        port( INPUT1,INPUT2 : BIT; OUTPUT : out BIT);
end entity NMOS_OR;

entity NMOS_XOR is
        port( INPUT1,INPUT2 : BIT; OUTPUT : out BIT);
end entity NMOS_XOR;
```

with the corresponding architectures. Using direct instantiation, an adder may be modeled as:

```
entity NMOS_ADDER is
        port( INPUT1,INPUT2,CATTY_IN : in BIT; OUTPUT, CARRY_OUT : out BIT);
end entity NMOS_ADDER;

architecture STRUCT of NMOS_ADDER is
        signal S1,S2,S3,S4 : BIT;
begin
        XOR1 :  entity NMOS.NMOS_XOR(BEH)      port map (B,CIN,S1);
        XOR2 :  entity NMOS.NMOS_XOR(BEH)      port map (S1,A,OUTPUT);
        OR1 :   entity NMOS.NMOS_OR(BEH)       port map (A,B,S2);
```

```
        AND1 :  entity NMOS.NMOS_AND(BEH)      port map (CIN,S2,S3);
        AND2 :  entity NMOS.NMOS_AND(BEH)      port map (A,B,S4);
        OR2 :   entity NMOS.NMOS_OR(BEH)       port map (S3,S4,CARRY_OUT);
end architecture STRUCT;
```

But the problem is that if the same design is required in a new technology, there is no way to benefit from the netlist information contained in the architecture above! The only solution is to create cut/paste operations to replace the references to NMOS library components by references to the new library component. This may represent a huge overhead and a nonneglectable risk of error in the case of very complex devices. Obviously, this kind of approach does not fit with good methodology.

The natural VHDL solution to increase the reusability of such structural models is to gather all binding information in configuration declarations. For this, general-purpose packages can be written containing component declarations independent of any particular technology. Then the structural architecture written using these declarations includes only the netlist information. For any new technology, the binding of library entities to sockets is achieved by configuration declaration.

Given the package declaration:

```
package GATE_PKG is
        component AND_GATE is
            port( INPUT1,INPUT2 : BIT; OUTPUT : out BIT);
        end component AND_GATE;

        component OR_GATE is
            port( INPUT1,INPUT2 : BIT; OUTPUT : out BIT);
        end component OR_GATE;

        component XOR_GATE is
            port( INPUT1,INPUT2 : BIT; OUTPUT : out BIT);
        end component XOR_GATE;
end package GATE_PKG;
```

the structural description of the adder can be written

```
use GATE_PKG.all;
architecture STRUCT of ADDER is
        signal S1,S2,S3,S4 : BIT;
begin
        XOR1 :  XOR_GATE  port map (B,CIN,S1);
        XOR2 :  XOR_GATE  port map (S1,A,OUTPUT);
        OR1 :   OR_GATE   port map (A,B,S2);
        AND1 :  AND_GATE  port map (CIN,S2,S3);
        AND2 :  AND_GATE  port map (A,B,S4);
        OR2 :   OR_GATE   port map (S3,S4,CARRY_OUT);
end architecture STRUCT;
```

To obtain a model corresponding to library NMOS, all we have to do is to write the configuration declaration

```
library NMOS;
configuration NMOS_ADDER of ADDER is
        for STRUCT       -- for architecture STRUCT
            for all : XOR_GATE use entity NMOS.NMOS_XOR(BEH); end for;
            for all : OR_GATE use entity NMOS.NMOS_OR(BEH); end for;
            for all : AND_GATE use entity NMOS.NMOS_AND(BEH); end for;
        end for;
end configuration;
```

If a new technology (TTL) implementation is required, then another configuration declaration is sufficient:

```
library TTL;
configuration  TTL_ADDER of ADDER is
        for STRUCT       -- for architecture STRUCT
            for all : XOR_GATE use entity  TTL. TTL_XOR(BEH); end for;
            for all : OR_GATE use entity  TTL. TTL_OR(BEH); end for;
            for all : AND_GATE use entity  TTL. TTL_AND(BEH); end for;
        end for;
end configuration;
```

So the TTL_ADDER and NMOS_ADDER share the same netlist description. Direct instantiation is useful, as long as it is not used too often.

6. INCREMENTAL BINDING

LRM REFERENCES: LRM 5.2.1

6.1. BACKGROUND

VHDL'87 requires that port maps and generic maps be fully specified once, and only once, for any component instance, either with a configuration specification inside an architecture or a component configuration inside a configuration declaration. Once a component instance is fully bound by an explicit configuration specification, the same instance cannot be bound again in a configuration declaration.

This one-step configuration mechanism is too strict for common design methodologies. For example, models are often written with their timing characteristics specified as generic parameters, so that these models are independent from any given technology. Later in the design process, real timings such as backannotated timings can be passed to these generic parameters. But for such models described at a structural level, the choice of the subelements of a given structural architecture (i.e., of the design units to be bound to components) is usually made long before the introduction of backannotated timings (i.e., before they can be passed on to generic parameters). A common practice in VHDL is to bind components to design units with configuration specifications written into the architecture (where the components are instantiated). But then, the insertion of timing information in generic maps must be done in the same configuration specification, and not "later," i.e., in configuration declarations. Such a two-step binding specification is possible in VHDL'92, which introduces an *incremental binding* mechanism.

Another related issue concerning the binding specification is the *default binding* mechanism. The definition of default binding in VHDL'87 contains some ambiguities and omissions that have been clarified in VHDL'92.

Finally, other clarifications have been made regarding the *binding time*, i.e., the time (analysis or elaboration time) at which any binding information is processed and checked. Ambiguities in VHDL'87 have led to different implementation choices resulting in different dependency rules between VHDL design units.

6.2. DESCRIPTION

In VHDL'92, component instances can be bound to design units in an incremental way; that is, the binding information can be split between a configuration specification (written in some architecture) and a configuration declaration. For this purpose, the syntax of the component configuration (written in a configuration declaration) has been changed to:

```
component_configuration ::=   for component_specification
                                  [ binding_indication ; ]
                                  [ block_configuration ]
                              end for ;

binding_indication ::=     [ use entity_aspect ]  [ generic_map_aspect ] [ port_map_aspect ]
```

The syntax of the configuration specification (written in an architecture or a block declarative part) also becomes

```
configuration_specification ::=  for component_specification  binding_indication ;
```

If the binding indication appears in a configuration specification, then the entity aspect is compulsory. On the other hand, the entity aspect can be omitted in the binding indication of a component configuration if all component instances listed in the component specification of this component configuration are already configured in one or more configuration specifications. More precisely, the block configuration

```
for SOME_ARCHITECTURE                          -- block configuration
      for L1, L2, L3 : SOME_COMPONENT          -- component configuration
            generic map ( ... ) port map ( ... ) ;  -- binding indication without any entity aspect
      end for ;
end for ;
```

is correct if and only if one or more configuration specifications written in the architecture SOME_ARCHITECTURE configure the instances L1, L2, and L3 of the component SOME_COMPONENT. For example:

```
architecture SOME_ARCHITECTURE of SOME_ENTITY is
      component SOME_COMPONENT ... end component ;
      for L1, L2 : SOME_COMPONENT use entity WORK.ENT(ARC1) ;
      for L3 : SOME_COMPONENT use entity WORK.ENT(ARC2) ;
begin
      ...
end architecture SOME_ARCHITECTURE ;
```

The binding indications written in the previous configuration specifications (which include an entity aspect) are called *primary binding indications*. The binding indications written in the previous block configuration (which have no entity aspect, but only a generic map and/or a port map) are called *incremental binding indications*: they are used to incrementally bind or rebind generics and ports.

Incremental binding is thus only possible between a configuration specification (written in some architecture or block) and a configuration declaration: the former contains the primary binding indications (which have compulsory entity aspect), and the latter contains the incremental binding indications (which are written inside component configurations).

Actually, two new mechanisms are made available to the designer:

- generics that have already been associated with actual values in a generic map of a configuration specification can be associated to other (possibly different) values with a generic map written in a configuration declaration, the latter generic map overriding the former; and
- ports that have *not* been associated with actual ports in a port map of a configuration specification (i.e., that have been associated with the **open** reserved word) can be bound later with another port map written in a configuration declaration.

For example, the following design units can be compiled in the given order in the working library:

```
entity AND3_ENT is
        generic ( DELAY : TIME );
        port ( A, B, C : in BIT; S : out BIT );
end AND3_ENT;

architecture DATAFLOW of AND3_ENT is
begin
        S <= A and B and C after DELAY;
end DATAFLOW;

entity EXAMPLE is
end;

architecture STRUCTURE of EXAMPLE is
        component AND2_COMP is
            port ( X, Y : in BIT; Z : out BIT );
        end component;
        for CI : AND2_COMP  use entity WORK.AND3_ENT(DATAFLOW )
                            generic map (DELAY => 2.0 ns)
                            port map (A => X, B => Y, C => open, S => Z);
        signal S1, S2, S3 : BIT;
begin
        CI :  AND2_COMP port map (S1, S2, S3);
end STRUCTURE;

configuration INCREMENTAL of EXAMPLE is
        for STRUCTURE
            for CI : AND2_COMP
                generic map (DELAY => 1.83 ns)     -- overriding the previous delay of 2.0 ns
                port map (C => '1' );              -- port previously left unconnected (open)
            end for;
        end for;
end INCREMENTAL;
```

6.3. USING THE FEATURE

When propagation delays are specified as generic parameters having default values, these delays are set to the default values if no configuration is given. For example, if the previous entity AND3_ENT is redeclared as follows:

```
entity AND3_ENT is
        generic ( DELAY : TIME := 3 ns );  -- default value for the generic parameter
        port ( A, B, C : in BIT; S : out BIT );
end AND3_ENT;
```

and the architecture STRUCTURE is rewritten using a configuration specification without a generic map:

```
architecture STRUCTURE of EXAMPLE is
        component AND2_COMPis
            port ( X, Y : in BIT; Z : out BIT );
        end component;
        for CI : AND2_COMP   use entity WORK.AND3_ENT(DATAFLOW )
                                              port map (A => X, B => Y, C => open, S => Z);
        signal S1, S2, S3 : BIT;
begin
        CI :  AND2_COMP port map (S1, S2, S3);
end STRUCTURE;
```

then the instance CI of component AND2_COMP is linked to the entity AND3_ENT with a generic parameter N set to its default value, i.e., 3 ns.

A further configuration declaration may then specify an explicit generic map with another time value (extracted by some backannotation tool, for example), thus allowing a kind of incremental binding in VHDL'87.

Alas, using the default value of generic parameters for this purpose is rather limited, particularly because all instances configured with the same entity (here, AND3_ENT) will use the *same* unique default value specified in this entity.

In a typical design methodology, timing characteristics are inserted in two or three steps:

- first, all timings are set to zero or to some unit delay (1 ns, for example), since the purpose of a simulation using such timings is just to check the functionality of the architecture (this step may be skipped and the next one directly used);
- then timings are set to estimated values depending on the *logic network* of the device: the propagation delay through some gate will depend on the number of devices driven by the output of this gate; the propagation delay through some data path operator will depend on the size (the number of bits) of this operator; and so on;
- finally, backannotation tools may extract more accurate timings from a first version of the synthesized *layout*, taking into account, for example, the delays introduced by block routing.

If the first timing values can be set to the default values of the generic parameters (because they are all equal to some unit delay), it is clear that the second timing characteristics may be specific to each instance of a given entity, and thus cannot be set to the default values specified in this entity. And this fact is even more critical for the backannotated values.

There is obviously a need for two successive generic maps for each instance to allow efficient backannotation. As illustrated by the figure 6.1, the incremental binding (allowing multiple generic maps) makes such a backannotation mechanism easier.

With the methodology given above, the second kind of delay (depending on the logic network) may be specified with generic maps in configuration specifications directly written into each architecture. When more accurate timings can be obtained later in the design process from a first version of the layout, then these timing values can override the previous ones if placed in generic maps of component configurations (which are themselves written in configuration declarations). These configuration declarations that set the definitive timing parameters could even be generated by some backannotation tools, depending on the CAD system used.

Of course, timings are only one type of information that can be passed through generic parameters, and many other applications using incremental binding can be found.

6.4. RELATED TOPICS

6.4.1. Default Binding

The writing of configurations is also made easier in VHDL'92 by two clarifications that have been made on the default configuration mechanism.

First, the insertion of implicit configuration specifications (i.e., the fact that component instances may be implicitly configured with an entity having the same name as the component) does not require the writing of block configurations, and thus does not need any top-level configuration declaration for that purpose (as it was the case in some implementations of VHDL'87).

```
library SOME_LIB;
architecture STRUCTURE of EXAMPLE is
        component AND2_COMP is
            port ( X, Y : in BIT; Z : out BIT );
        end component;
        -- if there is a visible entity named AND2_COMP with the same generic parameters and
        -- ports as the component above, then the following configuration is implicit:
        -- for all : AND2_COMP use entity SOME_LIB.AND2_COMP (ARC_NAME);
        -- where ARC_NAME is the most recently compiled architecture of entity AND2_COMP
        signal S1, S2, S3 : BIT;
begin
        CI : AND2_COMP port map (S1, S2, S3);
end STRUCTURE;
```

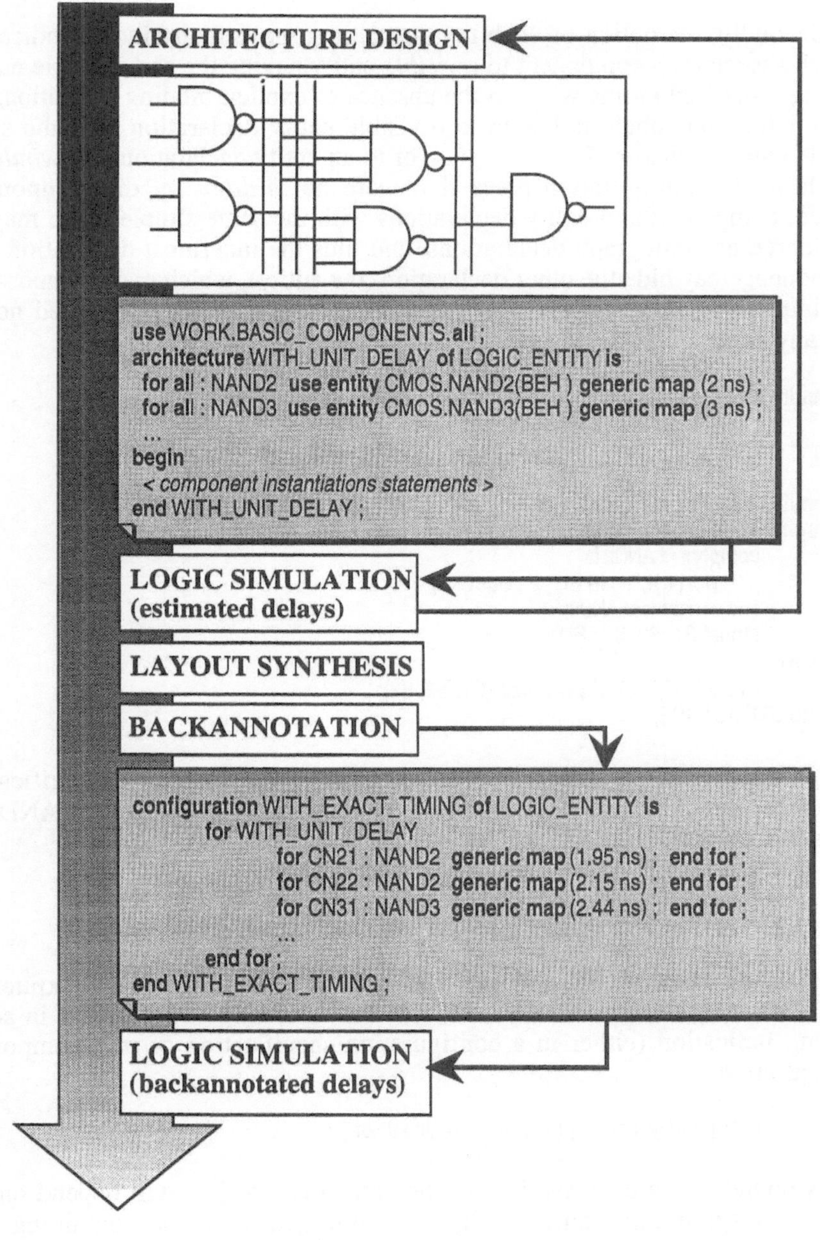

Fig 6.1 : Using Incremental Binding for Backannotation

Secondly, the notion of visibility concerning the default binding indication (which associates a component to a *visible* entity having the same simple name) has been clarified in this way: in the absence of explicit binding indication, the instantiated component is bound to a visible entity declaration with the same simple name as that of the component, or to an entity declaration that *would* be visible at the configuration place if it were not *hidden* by the component. Indeed, component and entity declarations with the same simple name may be considered as homograph declarations, and thus the innermost declaration (the component) may hide the other declaration (the entity), which makes impossible any implicit binding. In VHDL'92, the component name is guaranteed not to hide any entity of the same name.

```
entity AND2 is
        port ( X, Y : in BIT; Z : out BIT );
end entity AND2;

entity EXAMPLE is ... end;
architecture STRUCTURE of EXAMPLE is
        component AND2 is
            port ( X, Y : in BIT; Z : out BIT );
        end component AND2;
        signal S1, S2, S3 : BIT;
begin
        CI :  AND2_COMP port map (S1, S2, S3);
end STRUCTURE;
```

With the previous design units stored in the same library, this clarification guarantees that the entity AND2 will not be hidden by the component AND2 in the architecture STRUCTURE.

6.4.2. Binding Time

The time at which the binding indication is processed is made explicit in VHDL'92. Any design unit that contains the following entity aspect in some binding indication (either in a configuration specification or in a component configuration)

```
entity entity_name  [ ( architecture_identifier ) ]
```

is bound at the analysis time to the named entity, i.e., will depend on the named entity; if this named entity is recompiled, then the containing unit becomes obsolete and will have to be recompiled.

If this entity aspect effectively contains an architecture identifier, then the containing unit is also bound to the named architecture *if* this entity aspect is part of the binding indication of a component configuration that also contains an explicit block configuration, i.e.,

```
for component_instances : component_name                -- component configuration including :
        use entity entity_name (architecture_identifier) ;  -- a binding indication;
        for architecture_identifier ... end for ;       -- a block configuration.
end for ;
-- any design unit containing this component configuration is also bound to
-- the architecture named architecture_identifier
```

The same rule applies for an entity aspect with a configuration name:

```
for component_instances : component_name                -- component configuration including :
        use configuration configuration_identifier ;    -- a binding indication;
        for architecture_identifier ... end for ;       -- a block configuration.
end for ;
-- any design unit containing this component configuration is also bound to
-- the configuration named configuration_identifier
```

No additional analysis-time bindings can result from binding indications. Following are some typical examples of binding indications: for each of them, it is stated whether such a binding does or does not imply dependencies between design units.

```
architecture ARC1 of ENT1 is
        component COMP ... end component;
        for all : COMP use entity WORK.ENT2;            -- ARC1 depends on:
begin                                                   -- ENT1 & ENT2
end ARC1;

architecture ARC1 of ENT1 is
        component COMP ... end component;
        for all : COMP use entity WORK.ENT2(ARC2);      -- ARC1 depends on:
begin                                                   -- ENT1 & ENT2 (not ARC2)
end ARC1;

configuration CONF1 of ENT1 is
        for ARC1
            for all : COMP use entity WORK.ENT2(ARC2); -- CONF1 depends on:
            end for;                                -- ENT1, ARC1 & ENT2 (not ARC2)
        end for;
end CONF1;
```

```
configuration CONF1 of ENT1 is
      for ARC1
          for all : COMP use entity WORK.ENT2(ARC2);      -- CONF1 depends on:
              for ARC2 ... end for;                        -- ENT1, ARC1, ENT2 & ARC2
          end for;
      end for;
end CONF1;

entity COMP1 is end;
entity COMP2 is end;
architecture ARC1 of ENT1 is
      component COMP1 ... end component;
      component COMP2 ... end component;
      for all : COMP2 use entity WORK.COMP2;          -- ARC1 depends on:
begin                                                  -- ENT1 & COMP2 (not COMP1)
end ARC1;
```

When the containing unit is said to be bound to a named entity or architecture at the analysis time, it means that this containing unit depends on the named units, but also that all checks that can be performed at this time are effectively processed. For example, in the first four units listed above, this means that the port number, type, and mode conformance checking is performed between component COMP and entity ENT2 at the analysis time.

7. GROUPS

LRM REFERENCES: 4.6

7.1. BACKGROUND

By means of the user-defined attribute, VHDL allows the user to annotate his or her description with information that can be extracted and processed by tools. In synthesis for example, user-defined attributes have been successfully used to convey timing and area constraints to tools. In VHDL'92, this capability is extended by allowing all statements to be labeled and thus annotated. Still, user-defined attributes apply only to single entities[9] and not to groups of related items. So they fail to express in a clean way information such as pin-to-pin delay constraints.

7.2. DESCRIPTION

That is why VHDL'92 introduces the concept of a group to allow for the annotation of many related items with a single value. A group template is declared by a statement such as

[9] The name "entity" has several meanings in VHDL. Here it is used in the sense of "entity-class" as defined in the LRM (syntax summary): it is a named entity of the VHDL language, such as function, label, block, etc.

```
group PATH is (signal, signal); -- group template declaration
```

where PATH identifies the group template defined by the statement and enclosed in parentheses, and the reserved word **signal** means that PATH groups must be composed of two signal objects.

A group can be declared from the previously defined group templates PATH as follows:

```
group A_TO_S : PATH (A,S);
```

assuming that A and S refer to signal objects. A_TO_S can now be used to associate information that concerns A and S simultaneously. For example if we want to express a propagation delay between the two, we may write

```
attribute PROPAGATION_DELAY of A_TO_S : group is 250 ns;
```

assuming the attribute declaration

```
attribute PROPAGATION_DELAY : TIME;
```

The number of items in a group may be unlimited, and all the items need not belong to the same class. Of course, one given item may belong to several groups.

7.2.1. Syntax

The complete syntax of a group declaration and a group specification is given below.

```
group_template_declaration ::= group identifier is (entity_class [<>] { ,entity_class [<>] }) ;

entity_class ::= entity | architecture | configuration | procedure | function | package | type |
                 subtype | constant | signal | variable | component | label | literal | units | group | file

group_declaration ::= group identifier : group_type_name ( constituent {, constituent } ) ;
```

Both group template declaration and group declaration start with the reserved word **group** followed by an identifier designating the name of the declared group template in the first case, and the name of the group in the second.

A group template declaration consists in a list of one or many entity class names separated by commas. Groups of this type have the same number of members as the corresponding entity class. The box sign <> is optionally used to indicate different numbers of members belonging to the specified entity class.

Notice that a group may be a member of another group, because the reserved word **group** is allowed as an entity class.

A group declaration always refers to a group template declaration by its name. Following this information, the members of this group are listed. They must have consistent class corresponding to the entity classes listed in the group type declaration.

7.3. USING THE FEATURE

7.3.1. Timing Constraint

Given the entity declaration

```
entity CHIP is
        port(INPUT1,INPUT2 : BIT; OUTPUT1,OUTPUT2 : out BIT);
end CHIP;
```

let us consider the problem of expressing that the maximum propagation delay from any input to any output is 500 ns. We can state this by using the PATH group template and the PROPAGATION_DELAY attribute declarations defined earlier as

```
group INPUT1_TO_OUTPUT1 : PATH(INPUT1,OUTPUT1);
group INPUT2_TO_OUTPUT1 : PATH(INPUT2,OUTPUT1);
group INPUT1_TO_OUTPUT2 : PATH(INPUT1,OUTPUT2);
group INPUT2_TO_OUTPUT2 : PATH(INPUT2,OUTPUT2);
constant DELAY : TIME := 500 ns;
attribute PROPAGATION_DELAY of INPUT1_TO_OUTPUT1 : group is DELAY
attribute INPUT2_TO_OUTPUT1 of INPUT1_TO_OUTPUT2 : group is DELAY;
attribute PROPAGATION_DELAY of INPUT1_TO_OUTPUT1 : group is DELAY
attribute PROPAGATION_DELAY of INPUT2_TO_OUTPUT2 : group is DELAY;
```

This is correct but not optimal because the number of groups considered corresponds to the number of inputs multiplied by the number of outputs. The attribute specifications can be factored by the single specification:

```
attribute PROPAGATION_DELAY of all : group is DELAY;
```

which must still hold after the last group declaration provided no other group declaration follows!

Another more efficient way to do this is to define two group templates as follows:

```
group PIN_SET is (signal <>);
group PATHS is (group <>);
```

The box (<>) means one or many occurrences of items of the entity class preceding the <>. Thus the first declaration defines groups comprising one or many signals and the second defines groups of other groups.

Now we can declare two PIN_SETs as

```
group SOURCES : PIN_SET(INPUT1,INPUT2);
group TARGETS : PIN_SET(OUTPUT1,OUTPUT2);
```

and finally the timing constraint:

```
group INPUTS_TO_OUTPUTS : PATHS(SOURCES,TARGETS);
attribute PROPAGATION_DELAY of INPUTS_TO_OUTPUTS : group is DELAY;
```

Another example of timing specification/documentation is provided by the example of the greatest common divisor, presented in the next section. If we want the GCD computation to be executed in bounded time, we can express it as

```
group SEQ_PATH is (label, label);
attribute EXECUTION_DELAY : TIME;
group GCD_PATH : SEQ_PATH(START_STAT, LAST_STAT);
attribute EXECUTION_DELAY of GCD_PATH : group is 1000 ms;
```

7.3.2. Resource Allocation

Another possible application is to express resource sharing in a VHDL specification for synthesis. The idea is to define groups comprising statement labels and to specify resource constraints on these groups through attribute specification.

Consider the following specification of the greatest common divisor:

```
entity GCD is
        port(START : BIT; INPUT1,INPUT2 : INTEGER; OUTPUT : out INTEGER);
end GCD;
```

```
architecture BEHAVIORAL of GCD is
begin
	process
		variable A,B,R : INTEGER;
	begin
		wait until START='1';
		START_STAT : A := INPUT1;
		B := INPUT2;
		while A /= B loop
			if(A>B) then
				DIFF1 :   A := A-B;
			else
				DIFF2 :   B := B-A;
			end if;
		end loop;
		LAST_STAT : OUTPUT <= A;
	end process;
end BEHAVIORAL;
```

We can constrain the two minus operations to use the same functional unit "SYNTLIB.MINUS" using the attribute declaration

```
attribute IMPLEMENTATION : STRING;
```

and the attribute specification

```
attribute IMPLEMENTATION of DIFF1, DIFF2 : label is "SYNTLIB.MINUS";
```

Another way to specify such a constraint is to use group template declaration and group declaration:

```
group STATEMENTS is (label <>);
group MINUS_OP : STATEMENTS(DIFF1,DIFF2);
```

with the attribute specification

```
attribute IMPLEMENTATION of MINUS_OP : group is "SYNTLIB.MINUS";
```

If the desire is to express links between different items regardless of the value of the link, then the group concept can be used. For example, to specify that DIFF1 and DIFF2 should be performed on the same functional unit regardless of the identity of that unit, using the group template declaration

```
group SAME_FUNCTIONAL_UNIT  is (label <>);
```

we can write

```
group SAME_DIFF_OP : SAME_FUNCTIONAL_UNIT(DIFF1,DIFF2);
```

Introduction
New Simulation Mechanisms
New Structuring Mechanisms
=> ***New Interfacing Mechanisms***
New Predefined Operators, Functions & Attributes
Slight Enhancements
Language Simplifications
Clarifications
Annex

8. FOREIGN INTERFACES

LRM REFERENCES: 2.2

8.1. BACKGROUND

All computer languages need to define their interface with the external world. This external world is usually

- The host machine (memory management, disk accesses)
- The user (interactive IOs)
- Other pieces of software

Obviously, the first of these is unavoidable. Some system software can eliminate the user, but this is still impossible in VHDL. Communication with other pieces of software was not forbidden by VHDL'87 but was not really defined (one had to define a package declaration without a body, to be supplied by some *ad-hoc* mechanism - vendor dependent, if any). Moreover, only foreign subprograms could be handled this way (i.e., no entities or architectures). The only way to call foreign entities was by manipulating the **linkage** mode, with implementation-dependent meanings attached to it.

The only (pre)defined interface with the external world was, up to now, made by «magic» subprograms in VHDL (READ, WRITE, DEALLOCATE, etc.). Their semantics was given in somewhat unclear terms in the Reference Manual.

Of course, VHDL'92 does not remove those primitives, but rather enables an implementation to incorporate other foreign subprograms and other foreign entities[10]. It defines when the elaboration takes place or when an implementation is allowed to decide this.

8.2. DESCRIPTION

8.2.1. The Attribute FOREIGN

The FOREIGN attribute is defined in the package STANDARD (near the end):

```
package STANDARD is
        ...
        attribute FOREIGN : STRING;
end STANDARD;
```

8.2.2. FOREIGN Subprograms

Let us assume that our VHDL simulator is on a workstation with windows and menus. It might be nice control to a model through a menu-driven dialogue. Since the notion of a window is unknown in VHDL, we have to define a FOREIGN subprogram:

```
function HANDLEMENU(ARG : in MENU) return ITEM;
```

Then we have to "decorate" this declaration with the FOREIGN attribute:

```
attribute FOREIGN of HANDLEMENU : function is "W$MENU";
```

In this example, the name in the string is the external name: this line is the logical link between the internal name (HANDLEMENU) and the external name (W$MENU).

Of course, this W$MENU primitive has to be written somewhere in some language and then linked in some implementation-defined way.

The LRM does not specify what should appear in this string. Although it seems reasonable to think that the name of the called subprogram will be there,

[10] An entity is the couple entity declaration / one architecture.

any kind of information can be requested by a particular implementation: the number, the type of the arguments, the parameter passing conventions, etc.

Moreover, an implementation may place any restriction on the called subprogram, on its arguments, or even on its very existence: it is not mandatory for an implementation to deal with foreign subprograms.

8.2.3. Foreign Architectures

In the same way an external subprogram can be called via a foreign attribute, an entity declaration can be associated with a foreign architecture: the external view of the entity will be VHDL compliant, but the internal view will be elsewhere.

Again, an implementation is allowed to restrict the use of such foreign architecture and also not to allow them.

Foreign architectures can be used for two main goals: to interface with other simulators, and to protect the technology of the models.

Here is an example of a foreign architecture:

```
entity PROPRIETARY is
        generic (...);
        port (...);
end PROPRIETARY;

architecture LICENCED of PROPRIETARY is
        attribute FOREIGN  of PROPRIETARY : entity is "XYZ04/Licence:A2B8C4";
begin
end LICENCED;
```

8.3. USING THE FEATURE

8.3.1. Security

The foreign mechanism opens the door to all sorts of problems that VHDL is supposed to forbid by its construction. Foreign functions could wait, have side effects, modify their **in** arguments, etc. Instances of foreign architectures could communicate with each other using techniques hidden to VHDL; as in other better known domains, everything is a matter of moderation.

8.3.2. Do Not Look For It

It is not possible in VHDL to have foreign variables, signals, etc. - only subprograms and architectures. It is not possible (in the language, at least) to export subprograms or architectures.

9. READING & WRITING FILES

LRM REFERENCES: 4.3.1.4

9.1. BACKGROUND

Although many early implementations of subsets of VHDL simply ignored the input/outputs, more and more models today rely not only on their existence but also on their efficiency.

It may seem surprising that file input/outputs, which are based on a link with the file system of the host machine, might be an important part of a model that is supposed to describe some hardware.

There are actually two main uses for file input/outputs from within a model:

- To load random information in generic regular structures. A ROM or a PLA would be basically impossible to model in VHDL if it were not possible to describe a regular array of cells (most of the time generic in its sizes) and then to load their values from an external source of information. The only way to do this without files is to manipulate large aggregates in some package, which is quite a rigid and unfriendly process.
- To control, monitor, and profile a model. Mainly at the system level, the model has to be *observed* rather than *understood*. Input/outputs in the model can spy on or stimulate the model.

The problems in the VHDL'87 file input/outputs are of very different kinds:

- A number of errors or ambiguities have been found in the Reference Manual.
- The capabilities of VHDL in this domain were quite poor.
- The whole input/output system was irregular and designer hostile.

Typically, the above example of a ROM sized and loaded by the contents of a file is, according to the LRM, not feasible in VHDL'87. This is due to the definition of global staticness, which does not include user-defined function calls. However, global staticness is often *perceived* as "computable at elaboration time," a description that does include user-defined function calls.

```
G : for i in 1 to GET_NB_WORDS generate
        -- GET_NB_WORDS is a function reading from a file and returning a POSITIVE
        ...
end generate;
```

The result is that most implementations allow a user-defined function call (an input operation) to fix the genericity and to load the contents of a model. Some implementations are tied to the LRM and do not allow it. In this case, it can easily be proven that no means to workaround the problem exists: you simply cannot load the size and contents of a ROM from a file. However, if the size is fixed internally, the contents can be loaded into a signal; this means no loading before time 0, and also implies a signal without hardware meaning.

9.2. DESCRIPTION

There are now four classes of objects in VHDL: the FILE OBJECT is one of them. It was not clear in the LRM'87 how many classes of objects existed, and how much "variable" was a file.

As before, FILEs *cannot* be updated via assignments. They can be passed as parameters to a subprogram: a FILE is now a possible class in the parameter list:

```
procedure ADHOC(file F : TEXT);
```

Another source of misunderstanding was the mode of files for READ and WRITE predefined subprograms. A file parameter is now modeless, which shows that reading a file modifies the file (because a following read operation will not read the same thing), and, symmetrically, writing to a file implies reading something of it (e.g., its index).

9.2.1. Open and Close Operations

A major difference from VHDL'87 is that files can now be opened and closed (procedures FILE_OPEN and FILE_CLOSE). This is quite a change, since the lack of such primitives was intended and rationalized in VHDL'87: if one can open and close a file, one can communicate between processes without the use of signals. The implicitly defined procedures are given below.

Assuming the declaration

```
type TheFileType is file of TheTypeOfTheObject;
```

the following declarations are implicitly made just after the declaration of the file type:

```
procedure FILE_OPEN (file F : TheFileType;
                     EXTERNAL_NAME : in STRING;
                     OPEN_KIND : in FILE_OPEN_KIND := READ_MODE);
procedure FILE_OPEN (STATUS : out FILE_OPEN_STATUS;
                     file F : TheFileType;
                     EXTERNAL_NAME : in STRING;
                     OPEN_KIND : in FILE_OPEN_KIND := READ_MODE);
```

Note that the existence of a STATUS in this second FILE_OPEN permits for testing the existence of a file (Open the file, test the status, report if there is an error, else continue). This feature was completely lacking in VHDL'87.

```
procedure FILE_CLOSE (file F : TheFileType);
procedure WRITE (file F : TheFileType, VALUE : in TheTypeOfTheObject);
function ENDFILE (file F : TheFileType) return BOOLEAN;
```

Depending on *TheTypeOfTheObject*,

```
procedure READ(file F : TheFileType, VALUE : out TheTypeOfTheObject);
-- if the type of the object is constrained

procedure READ(file F : TheFileType,
               VALUE : out TheTypeOfTheObject;
               LENGTH : out NATURAL);
-- if the type of the object is not constrained (e.g., STRING or BIT_VECTOR).
```

9.2.2. Package TEXTIO

TEXTIO is a package that is part of the standard and deals with human-readable input/outputs.

The changes in TEXTIO are mainly corrections:

- The new functionalities are provided by the above-mentioned FILE_OPEN and FILE_CLOSE, which are automatically declared together with the type TEXT.
- The files INPUT and OUTPUT are now declared **open**.
- The file parameter of the procedures READLINE and WRITELINE is now **inout** for the reasons explained above.
- The function ENDFILE is now an impure function (see chapter 10). Indeed, the bug of VHDL'87 becomes a feature.

9.3. USING THE FEATURE

A very useful application of TEXTIO is to load from a file the size and the contents of a pseudo-regular structure, such as a ROM or a PLA.

For example, given the user-written file,

```
255
7
01001010
and 255 other values
...
```

we can write such a code to load

- an integer from the file
- a BIT_VECTOR from the file (note the passing of an unconstrained array as an **inout**); and
- finally, the whole file in an array.

```
file THE_FILE : TEXT open READ_OPEN is "TheExternalName";

impure function GET_INTEGER return INTEGER is
        variable THE_LINE : LINE;
        variable THE_INTEGER : INTEGER;
begin
        READLINE(THE_FILE, THE_LINE);
        READ(THE_LINE, THE_INTEGER);
        return THE_INTEGER;
end GET_INTEGER;
```

```
procedure GET_A_BIT_VECTOR (THE_BIT_VECTOR : inout BIT_VECTOR) is
        variable THE_LINE : LINE;
begin
        READLINE(THE_FILE, THE_LINE);
        READ(THE_LINE, THE_BIT_VECTOR );
end GET_A_BIT_VECTOR;

••• here unrelated code

constant ROM_SIZE : INTEGER := GET_AN_INTEGER;    -- GET_AN_INTEGER will return 255
constant WORD_SIZE  : INTEGER := GET_AN_INTEGER; -- GET_AN_INTEGER will return 7

••• here unrelated code

process
        type ROM_CONTENT is  array (0 to ROM_SIZE) of BIT_VECTOR(0 to WORD_SIZE);
        variable CONTENTS : ROM_CONTENT;
begin
        for I in 0 to ROM_SIZE loop
             GET_A_BIT_VECTOR(CONTENTS(I));
             -- GET_A_BIT_VECTOR is a procedure (here will load CONTENTS(I))
        end loop;
        ... -- should be complicated to take care of error cases (file too short, etc.).
end process;
```

10. IMPURE FUNCTIONS

LRM REFERENCES: 2

10.1. BACKGROUND

Functions in VHDL'87 have an original property among most known languages: they are pure mathematical functions, which means that they will return the same result for the same set of inputs, with the remarkable exception of functions calling NOW. The consequences are as follow.

- A function has only **in** parameters.
- The parameters of a function are of class **constant**. In particular, no access type, nor composite made of access types, can be allowed as an argument (access types are **variables**).
- A function cannot have side effects. This implies a complicated set of rules concerning the access to objects declared outside of the function.
  It happens that the set of rules defined for this purpose has no holes: it is *really* impossible in VHDL to have a function PULL returning the top of the stack *and* pulling the stack (a side effect). You also cannot write a function returning the next character from a file as in C or Pascal, because you will always get the same character: the first one! Note that, in the following example, the file could not be declared outside of the function (access to an external object).

```
function GET_CHAR return CHARACTER is
        file F : TEXT is in "TheFile";              -- the file is opened here
        variable C : CHARACTER;
begin
        READ(F,C);                  -- you always get the first character of the file
        return C;
end GET_CHAR;                       -- the file is closed here
```

This example shows also that the OPEN and CLOSE subprograms do not exist in VHDL'87: this was deliberate for the sake of determinism (no inter process communication without a signal) and purity of function (no fooling the checkings via files).
Although desirable in certain cases (in particular for type-conversion functions and for resolution functions), such an enforcement becomes hard to live with in other cases.
The system-level designers would like to have a random-number generator: this cannot be a function in VHDL'87 because it needs a "seed"[11] for its algorithm.
Reading a file is a common activity, but this is particularly tricky in VHDL.
Last but not least, if generic actuals have to be computed at elaboration time, they *have to* be computed by functions and *cannot* be computed only by procedures - because a value, not a statement is needed. Should such a generic depend on a value provided interactively or in a file, LRM'87 formally forbids it. Fortunately, most implementations allow it, but the purpose of restandardization was to fix this kind of discrepancy.

10.2. DESCRIPTION

VHDL now allows functions to be "impure," which means that they can have access to external data.

```
impure function F (V : INTEGER) return INTEGER;
```

Such a function can read an external file, have a "seed", be foreign (see chapter 8). However, in this last case, the purity check is left to the foreign part. Note that in any case the only mode for input parameters remains the **in** mode, and the actual cannot be (or include as a sub element) an access type.

---

[11] Obviously, random-number generators cannot be completely random: «Anyone who considers arithmetical methods of producing ramdom digits is, of course, in a state of sin» (Von Neumann as cited by Knuth). They compute so-called random numbers with polynomial algorithms, using shifts and XORs on a remanent memory. The seed is the memory of the random-generator.

For the sake of orthogonality, the **pure** reserved word has been added to the language and will be optional, but by default on "non-impure" functions.

A consistent set of rules defines where such a function can be called. Obviously, it cannot be called (even indirectly) from a pure function. It cannot be a resolution function. However, an impure function can be used to initialize a constant. Believe it or not, a non-static expression such as an impure function can be used to initialize a constant, which is a (globally) static value.

The purity of the function is not part of its profile, and no signature can distinguish a pure and an impure function.

Resolution functions must be pure.

10.3. USING THE FEATURE

Examples of the use of impure functions can be found in the chapter section 9.2.2) and in the chapter 4.

Introduction
New Simulation Mechanisms
New Structuring Mechanisms
New Interfacing Mechanisms
=> ***New Predefined Operators, Functions & Attributes***
Slight Enhancements
Language Simplifications
Clarifications
Annex

11. SHIFT AND ROTATE OPERATORS

LRM REFERENCES: 7.2

11.1. BACKGROUND

VHDL'87 did not define any shift or rotate operators. The main problem here was reaching a consensus on a minimum set of operators. This has now been done, and VHDL'92 predefines four shift and two rotate operators.

11.2. DESCRIPTION

The six shift and rotate operators are binary operators called **sll**, **srl**, **sla**, **sra**, **rol**, and **ror**. They are only defined for one-dimensional array types (vectors) whose element type is either BIT or BOOLEAN. The left operand must be of that type. The right operand is of type INTEGER:

- if this operand equals zero, no operation at all is performed: (A **sll** 0) is a null operation whose result is A.
- if the right operand is positive, then the shift or rotate operation is repeated this number of times: (A **sll** 5) performs a one-bit-shift five times.

• if the right operand is negative, then the opposite shift is performed a number of times corresponding to the absolute value of the right operand: (A **rol** -6) is equivalent to (A **ror** 6).

Of course, the main value of having integer as the right operand is the possibility of using an expression (A **ror** N*P/I) that is dynamically evaluated. In this case, the operator must be considered as a "generic" rotate operator (**ror** or **rol**) depending on the sign of the expression.

The semantics of the basic operation (one-bit rotate or shift) is illustrated in figure 11.1. Some of these operations (**sll, srl, sla, sra**) imply losing the value of one element. In some cases (**sll** and **srl**), a fill value is used. It is defined as the value of the attribute 'LEFT on an element type of the array: '0' is the fill value for BIT_VECTOR operations and FALSE is the fill value for boolean vectors. The only possibility of changing this fill value is to overload the operators.

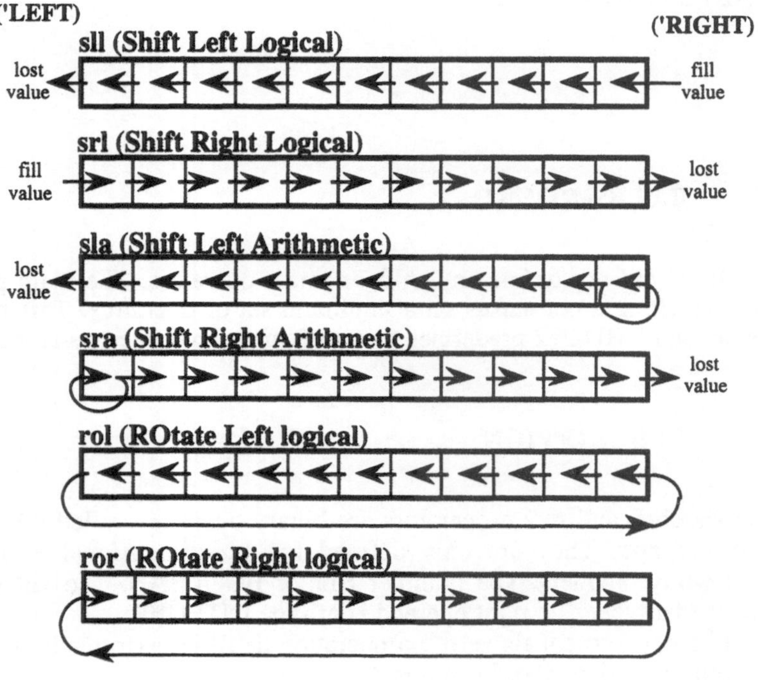

Fig 11.1 Shift and Rotate Operator Semantics

11.3. USING THE FEATURE

Since these six operators are a trade-off between the different needs of designers, we can imagine some cases where overloading will occur:

- In the package STD_LOGIC_1164, these operator will appear to perform their operations on STD_ULOGIC.
- From a system level point of view, it might seem frustrating to handle only vectors of bits or booleans. A simple overloading will allow us to redefine these operations on INTEGER (for example) with a fill value of zero (why not ?).
- At a lower level, the fill value for shifting arrays of bits can contradict the technology: the fill value must be '1' instead of '0'. Overloading will also solve this.
- For synthesis purposes, it can be difficult to map an operator such as (A **ror** N) to an existing device if N can be either positive or negative. An overloaded version restricted to positive right operands can be exported by a synthesis library in that case.

Indeed, the main purpose of these six new operators of VHDL'92 is to provide new reserved words and their semantics (which should be respected when overloaded).

12. XNOR OPERATOR

LRM REFERENCES: 7.2

12.1. BACKGROUND

A set of five logical operators is predefined in VHDL'87: **and, or, nand, nor**, and **xor**. For obvious reasons of symmetry, some users requested the addition of a sixth operator: **xnor**.

12.2. DESCRIPTION

The operator **xnor** has now been added to the language. Its semantics is clear: A **xnor** B is equal to **not** (A **xor** B). This operation is also called identity: the result is true if the values of the operands are the same. The operands, as is the case for the other logical operators, can be of type BIT, BOOLEAN, or a vector (single dimensional array) of such types (with the same length). The operands must have the same base type (BIT, BOOLEAN, vector of BIT, or vector of BOOLEAN). As shown in figure 12.1, the result is of the same base type.[12] Logical operators are the lower-priority predefined operators in VHDL (except for the unary logical operator **not**).

---

[12] This is the main difference with the use of the "=" operator, which would always return a boolean value.

Base type of A and B is BIT

A	B	A xnor B
0	0	1
0	1	0
1	0	0
1	1	1

Base type of A is BIT and base type of B is BOOLEAN

A	B	A xnor B

not allowed

Base type of A and B is BOOLEAN

A	B	A xnor B
FALSE	FALSE	TRUE
FALSE	TRUE	FALSE
TRUE	FALSE	FALSE
TRUE	TRUE	TRUE

Fig 12.1 Xnor Operator Truth Tables

12.3. USING THE FEATURE

This feature is straightforward to use. Nevertheless, its introduction provides an opportunity to point out some supposedly well-known mechanisms of VHDL('87):

- It is possible to overload this operator, and this will be done in the 1164 package (actually it is already in the package, but commented out).
- When handling **or**, **and**, **nand**, or **nor** operators, it may be possible in some cases to predict the result once the first operand value is known: this is the notion of "short-circuit" operation. For example, if the first operand of an **and** operator is FALSE, the result is false whatever the value of the second operator. It is important to know that VHDL does not evaluate the second operand in such cases. This allows us to use expressions such as: **if** (A=0 **or** B/A=1). This "non-evaluation" can lead to treacherous behavior of the program: when writing A **and** F(B), an assert statement (for example) within the function F will be executed (even if the assert condition is false) only if A is TRUE. A quick glance at the truth table shows that the two operands of the **xnor** operator, as is the case for the **xor** operator, have to be evaluated to compute the result: **xnor** is not a short-circuit operation.
- The **xnor** operation is not associative. This property has two consequences:

- The syntax for a VHDL expression does not allow a sequence of **xnor** operators. A **xnor** B **xnor** C is prohibited and should be rewritten with parentheses. The same expression with **and** or **or** operators is legal because these operators are associative.

- This operator should be carefully used in a resolution function, since the user does not handle the manner in which the single input parameter used by such a function represents the multiple sources. The order of this collection of inputs is implementation dependent. This implies that a resolution function whose result depends on the order of the parameters can yield different results even if called with the same set of input values. This breaks the determinism of VHDL and produces non-predictable and non-portable models.

Since **xnor** is a new reserved word, using XNOR as an identifier is no longer allowed. If such an identifier has already been used in a previous VHDL'87 source code, it is a good practice to change it to an extended identifier: \XNOR\. This new VHDL'92 feature is described in chapter 22 (note that this kind of identifier is case sensitive). This change will make the code compliant with the VHDL'92 standard without modifying its readability (i.e., its documentation power).

13. PREDEFINED ATTRIBUTE DRIVING_VALUE (AND 'DRIVING)

LRM REFERENCES: 14

13.1. BACKGROUND

Especially when modeling simple hardware devices, it can seem very restricting not to be allowed to read output port values (i.e., to place them on the right-hand side of assignment statements, as shown in figure 13.1).

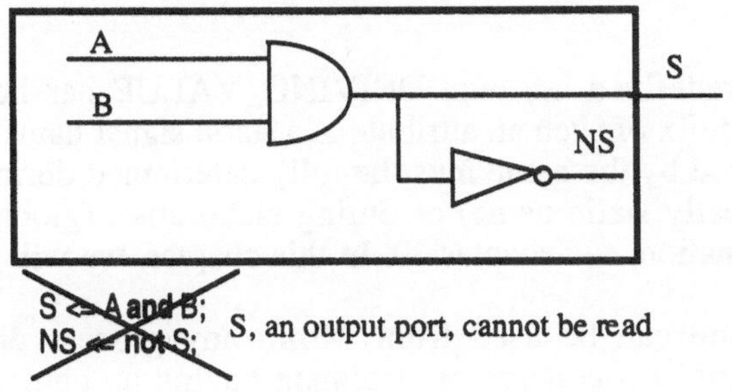

Fig 13.1 No Reading Output Ports

What is the fundamental reason for this restriction on the use of the output port value?

A resolution function can be applied to an output signal. In this case, reading the signal value means reading the resolved value and not only the contributing value of the component (in the example of figure 13.2). In other words, this is a way to enter information (the resolved value depends on other signal values in the network) using an output port. The value S' of figure 13.2 can be very different from the value S".

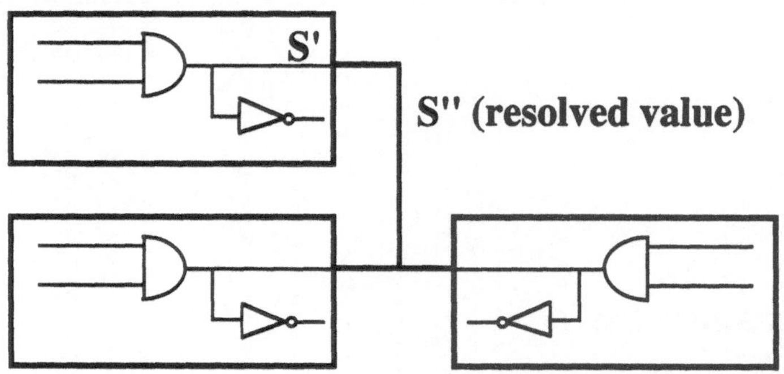

Fig 13.2 Reasons for a Restriction

But what is the real need of the designer? He or she usually only wants to read the contributing value S' (i.e., the value he or she just assigned to the port before resolution). VHDL'92 gives the designer this ability while respecting VHDL consistency.

13.2. DESCRIPTION

The new predefined attribute DRIVING_VALUE has been added to the language. The prefix of such an attribute is a static signal name. This means that the object denoted by the name must be fully determined during analysis of the VHDL text (locally static name) or during elaboration (globally static name). For more information, see chapter 30. In this chapter, we will use S as a signal name.

This attribute can be used from within any process driving S, (i.e., a process statement or a concurrent statement having an equivalent process) or from any procedure with S as an **out** or **inout** signal parameter.

As shown in figure 13.3, this attribute returns the contributing value of the process (or the procedure) to the signal S. It is important to note that the result of this attribute is a value and not a signal.

An error will occur if the corresponding process does not drive the signal S because 1) there is no driver or 2) because this driver has been disconnected (i.e., the driver is determined by the null transaction). A static verification of the source code can determine that there is no driver (case 1), and a new predefined attribute 'DRIVING gives the dynamic information of disconnection (case 2).

The 'DRIVING attribute returns a boolean value that is FALSE if the driver has been disconnected. It can be used from within any process driving S or from any procedure with S as an **out** or **inout** signal parameter. However its use is uncommon. When writing a process, a designer usually knows if the signal is disconnected or not. Nevertheless, when the disconnection depends on a complex condition or is related to a given time (disconnect clause), the attribute 'DRIVING may be used in the following manner to avoid an error when calling 'DRIVING_VALUE:

```
if S'DRIVING then
        V := S'DRIVING_VALUE;
end if;
```

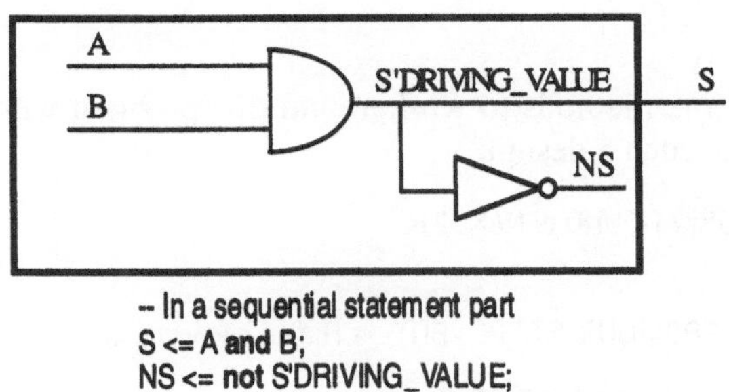

Fig 13.3 S and S'DRIVING_VALUE

13.3. USING THE FEATURE

13.3.1. An Example

The attribute 'DRIVING_VALUE will no doubt become popular among designers. There are many simple cases in which it is useful to refer to the contributing value of the output port.

The following example describes a nand gate whose propagation delay depends on the edge.

```
entity NAND2 is
        generic (FROM_0_TO_1, FROM_1_TO_0 : TIME);
        port (A, B : in BIT; OUTPUT : out BIT);
end NAND2;

architecture VHDL92 of NAND2 is
begin
process
begin
if (A nand B) /= OUTPUT'DRIVING_VALUE then
        -- (A nand B) /= OUTPUT would be incorrect.
        if OUTPUT'DRIVING_VALUE = '0' then  --    OUTPUT = '0' would also be incorrect.
            OUTPUT <= A nand B after FROM_0_TO_1;
        else
            OUTPUT <= A nand B after FROM_1_TO_0;
        end if;
end if;
wait on A,B;
end process;
end VHDL92;
```

Of course, some methods to workaround this problem were well-known in VHDL'87 to make such a design:

```
architecture WORKAROUND of NAND2 is
begin
process
        variable PREVIOUS_STATE : BIT;  -- The ad-hoc variable
begin
if (A nand B) /= PREVIOUS_STATE then
        if PREVIOUS_STATE='0' then
            OUTPUT <= A nand B after FROM_0_TO_1;
        else
            OUTPUT <= A nand B after FROM_1_TO_0;
        end if;
        PREVIOUS_STATE := A nand B;
end if;
wait on A,B;
end process;
end WORKAROUND;
```

But creating an extra variable, local to the process, or an extra signal, local to the architecture, complexifies the task of the reader. A human has to understand the use of this object, and a synthesis tool, which is another example of a possible reader, has to ignore it. The readability of source code is improved by using the new attribute 'DRIVING_VALUE.

13.3.2. Another Example[13]

The attributes 'DRIVING and 'DRIVING_VALUE are also useful when the contribution to the signal comes from the body of a procedure, which is not visible (or maybe not written yet).

In this example, a signal S is assigned, then passed as an argument to the procedure PROC. It is possible, after the call, to determine if the signal is still driven (in case of a disconnection, and then a wait statement greater than 3 ns in the procedure), and what is the driving value:

```
procedure PROC (signal Arg : out BIT_RESOLVED) is
...
end PROC;
...
signal S : BIT_RESOLVED register;
disconnect S: BIT_RESOLVED after 3 ns;
...
LBL : block (...)
begin
...
        process
        begin
            ...
            if GUARD
                S <= '1';
                PROC(S);
                if S'DRIVING then                     -- S is still driven
                    ... use of S'DRIVING_VALUE       -- S'DRIVING_VALUE is not S!
                else                                  -- S is no more driven
                    ...
                end if;
            else
                ...
            end if;
        end process;
...
end block;
```

13.3.3. Limit of 'DRIVING_VALUE

A "limit" of the use of 'DRIVING_VALUE is that the returned value is only that of the process in which the attribute is called and the signal is driven.

If the attribute were called in a process (or equivalent process) where no driver existed for a signal, an error would be flagged by the compiler.

---

[13] Thanks to Oz Levia.

For example, it might be useful in an RS gate to be able to read the 'DRIVING_VALUE of a driver outside the concurrent statement (the process in which the attribute is called. Unfortunately, this is not possible.

14. PREDEFINED ATTRIBUTE 'ASCENDING

LRM REFERENCES: 14

14.1. BACKGROUND

In some cases, the logic of operations performed on an array depends on its direction: **to** or **downto**. If this direction is not known in advance, e.g. because this array is declared as an unconstrained formal parameter of a subprogram, the predefined attributes LEFT, RIGHT, HIGH, and LOW can be used.

Indeed, comparing the value of the attribute HIGH to the attribute LEFT is not enough to determine the direction. If the attribute HIGH is equal to the attribute RIGHT, this does not mean anything about the direction: the array may have a single element or may be a null array. Even if the attribute LEFT is greater than the attribute RIGHT, it might be the pathological case:

```
type FOO is array (10 to 1) of WHATEVER; -- null array
```

14.2. DESCRIPTION

VHDL'92 provides a new predefined attribute: 'ASCENDING. With reference to attributes 'LEFT or 'RIGHT, 'ASCENDING can be applied on any scalar type (or subtype) or array object. It returns a boolean value that is TRUE if the range is ascending, and false otherwise.

14.3. USING THE FEATURE

14.3.1. An Initial Example

The following function extends the most significant bit side (highest indice) of a given bit vector A up to a length L using the bit value B. In this example, the attribute 'ASCENDING allows us to handle the case of a single-element array and a null array. Without this attribute, the only strategy consists of forcing the range of such arrays to be ascending (or descending).

```
function EXTEND (A : in BIT_VECTOR; L : in NATURAL; B : BIT) return BIT_VECTOR is
        constant V_TO : BIT_VECTOR(1 to L) := (others => B);
        constant V_DOWNTO : BIT_VECTOR(L downto 1) := (others => B);
begin
if A'ASCENDING then
        V_TO (1 to A'LENGTH) := A;
        return V_TO;
else
        V_DOWNTO (A'LENGTH downto 1) := A;
        return V_DOWNTO;
end if;
end EXTEND;
```

14.3.2. Overloading of the VHDL'87 Concatenation Operator

The algorithm of the concatenation operator has changed in VHDL'92 (see chapter 28). This is a potential source of non-portability. The following source code can be used to emulate the "old" VHDL'87 concatenation operator:

```
function "&"(V1, V2 : BIT_VECTOR) return BIT_VECTOR is
        constant R_TO : BIT_VECTOR(V1'LEFT to V1'RIGHT + V2'LENGTH) ;
        constant R_DOWNTO : BIT_VECTOR(V1'LEFT downto V1'RIGHT - V2'LENGTH);
begin
if V1'ASCENDING then
        R_TO(V1'LEFT to V1'RIGHT) := V1 ;
        R_TO(V1'RIGHT+1 to R_TO'RIGHT) := V2 ;
        return R_TO ;
else
        R_DOWNTO(V1'LEFT downto V1'RIGHT) := V1 ;
        R_DOWNTO(V1'RIGHT-1 downto R_TO'RIGHT) := V2 ;
        return R_DOWNTO ;
end if ;
end;
```

15. PREDEFINED ATTRIBUTES 'BEHAVIOR & 'STRUCTURE

LRM REFERENCES: 14

15.1. BACKGROUND

The definitions of these two VHDL'87 attributes have been discussed extensively. The final conclusion was that if the definitions were not adapted, the attributes themselves would not correspond to a real need of the designer. In other words, nobody used these attributes.

15.2. DESCRIPTION

As a consequence of and in spite of the upward-compatibility design objective, these two predefined attributes have been removed from the VHDL'92 standard.

15.3. USING THE (ABSENCE OF) FEATURE

The "behavioral" or "structural" nature of some parts of the design is a characteristic that can be evaluated statically just by looking at the source code.

If some applications need such features, user-defined attributes can be declared and used for this purpose.

16. PREDEFINED ATTRIBUTES 'IMAGE & 'VALUE

LRM REFERENCES: 14.1

16.1. BACKGROUND

It is often desirable to convert a value of some type that has a clear and non ambiguous representation into a value of type STRING: for example, to display the value of an enumeration type in an assertion statement.

In the same way, when acquiring data, it should be useful to convert STRING type values into some scalar types.

16.2. DESCRIPTION

VHDL'92 offers a two-way conversion of a string into any scalar type or subtype (integers, floating types, physical types, and enumeration types). This can be done using two new predefined attributes - 'IMAGE and 'VALUE - that work on types and return a value.

```
scalar_type_or_subtype'IMAGE(value_of_this_type)
```

returns a string representing the scalar type value.

```
scalar_type_or_subtype'VALUE(string)
```

returns a value of the given scalar type.

Remarks: The attribute 'VALUE is case insensitive, but an error occurs if the string parameter does not correspond to the string representation of any of the values of the type or subtype denoted by the prefix. The expression BOOLEAN'VALUE("True") is legal, but the expression BOOLEAN'VALUE("Unknown") will generate an error because UNKNOWN is not an element of the enumerated type BOOLEAN.

16.3. USING THE FEATURE

16.3.1. Displaying Values

A very common and very useful use of the 'IMAGE attribute is the conversion of a scalar type into a string to constitute a message.
The code

```
        ...
        variable N : INTEGER;
begin
        ...
        for I in 1 to N loop
            COMPUTE(I);
            report "End of computation number " & INTEGER'IMAGE(I);
        end loop;
        ...
```

displays messages of the form

```
End of computation number 1
End of computation number 2
etc.
```

16.3.2. Reading and Writing ASCII Files

Reading and writing ASCII files is very popular in VHDL. ASCII files are a way to store sequences of stimuli and to change them without recompilation. They are also a way to trace the behavior of a program.

The standard package TEXTIO of the library STD offers read and write primitives on ASCII files. It is therefore possible to read types INTEGER, BOOLEAN, BIT, etc. Unfortunately, it is not possible to operate on a user-defined enumerated type. The use of 'IMAGE and 'VALUE solves this. The following example shows how to write a package exporting primitives to read and write ASCII files containing one BOOLEAN value (predefined enumerated type) and a user-defined enumerated type: MY_ENUMERATED_TYPE.

```
package MY_TYPES is
        type MY_ENUMERATED_TYPE is (ONE, TWO, THREE);
end package;

use TEXTIO.all;
use WORK.MY_TYPES.all;
package body STIMULIS is

        procedure READ_VALUE (F : inout TEXT; VAL : out BOOLEAN ) is
            variable L : LINE;
        begin
            READLINE(F, L);
            READ(L,  VAL);
        end READ_VALUE;

        procedure READ_VALUE (F : inout TEXT; VAL : out MY_ENUMERATED_TYPE ) is
            variable L : LINE;
            variable S : STRING(1 to MAX);
            variable LL : INTEGER;
        begin
            READLINE(F, L);
            LL := L'LENGTH;
            READ (L, S(1 to LL));
            VALUE  := MY_ENUMERATED_TYPE'VALUE(S(1  to LL));
        end READ_VALUE;

        procedure WRITE_VALUE (F : inout TEXT; VAL : in BOOLEAN ) is
            variable L : LINE;
        begin
            WRITE(L, VAL);  -- This is equivalent to write(L, INTEGER'IMAGE(VAL));
            WRITELINE(F, L);
        end WRITE_VALUE;

        procedure WRITE_STIMULI(F : inout TEXT; VALUE : in MY_ENUMERATED_TYPE )is
            variable L : LINE;
        begin
            WRITE  (L, MY_ENUMERATED_TYPE'IMAGE(VAL));
            WRITELINE(F, L);
        end WRITE_VALUE;

end;
```

Of course, reading or writing a predefined enumerated type (BIT, BOOLEAN) was already possible in VHDL'87 but the same functionality on a user-defined enumerated type was very cumbersome to write. For example, a read operation implies reading a string, putting it in upper case, and comparing it to all possible my_enumerated_type'VALUE(enumerated_type_values) in order to return the right result. Therefore, the attributes 'IMAGE and 'VALUE represent real progress.

17. ATTRIBUTES 'PATH_NAME 'INSTANCE_NAME 'SIMPLE_NAME

LRM REFERENCES: 14.1

17.1. BACKGROUND

It is often the case that, during the simulation, an error is difficult to locate because it happens to be in one instance of a component instantiated many times. In this case, the classical assert statement of VHDL'87,

```
assert Condition report "Error in Entity E(A)";
```

can identify which entity was bound on a given component, but nothing about the component instance itself.

17.2. DESCRIPTION

The attribute 'SIMPLE_NAME provides a simple way to know from within the model the name of a named entity[14] (i.e., nearly everything that has a

[14] The word **entity** has several meanings in VHDL. Here it has more or less the meaning of "concept". See the rule *entity_class* in appendix A of the LRM. See also section 5.1 of the LRM.

name, as defined in section 5.1 of the LRM). For example, if V is a variable, V'SIMPLE_NAME returns the string "v". Note that the information returned is always in lower-case.

The attribute 'PATH_NAME, applicable to any named entity some of those elaborated at run-time (e.g., loop parameters, allocators), returns a string (lower case again) that reflects all labels in the path of this given name.

VHDL'92 also provides an attribute 'INSTANCE_NAME, applicable to any entity in VHDL (here, **entity** has the same meaning as before), with the same exceptions. This attribute returns a string (lowercase) that includes PATH_NAME and also the configuration information on each label of component.

Typically, it is now possible to write

```
report "Now in " & LBL'PATH_NAME;
```

where **report** is the new report statement (see chapter 27). LBL'PATH_NAME will return a string uniquely identifying the entity LBL (here, a label), in terms of its name after elaboration. All three of these attributes are globally static values.

The syntax obtained out of 'INSTANCE_NAME is a bit complicated: this is because the string includes information about both the design hierarchy (components and configurations) and the structure of the design unit (blocks and processes). Moreover, the leftmost information has to do with the name of the library itself, which is implementation dependent.

•'INSTANCE_NAME ONLY :

Within a string given by 'INSTANCE_NAME, the substring

```
...lbl@ent(arch)...
```

refers to a component labeled by LBL, and configured by the entity ENT associated with the architecture ARCH.

• 'PATH_NAME & 'INSTANCE_NAME

The following conventions apply to both 'INSTANCE_NAME and 'PATH_NAME.

The substring

```
...a:b:c:d
```

refers to a hierarchy of names within the same design unit (for example, variable D within process C within block B within block A).

The substring

```
...a:b(5):c...
```

refers to some label C within a generate statement whose index is 5, within a block A[15].

A full instance-name may be very long and unreadable, such as

```
:top(top):I2@bottom(bottomarch):processbottom:v
```

If the root of the path is a package, the name of the library of the package must prefix the string, and this name is the logical one. For example,

```
:library:pack:object
```

Here, the designation of the library (:library:) is the logical name of the library.

In case a process does not have a label, the substring

```
...a:b::c:d...
```

with the pattern '::' indicates that a label was not given to some process. Because of this, a 'PATH_NAME or 'INSTANCE_NAME may be not unique within a given design.

17.3. USING THE FEATURE

The main use of these attributes, from an end-user point of view, will be for error messages and reports. For example, the statement

```
report "I am now in " & MyLabel'PATH_NAME ;
```

might be very useful for debugging purposes. The result will be a report, i.e., a message on the screen, with the full hierarchical path name of the mentioned label.

These attributes might also be hooks for some tools: the use of 'INSTANCE_NAME may be envisaged for backannotation purposes, although it seems to be very inefficient if made in pure VHDL. Let us just mention that it is possible to store in a text file a list of path names associated with values

[15] Since the index of a generate statement can be a name (a constant or an enumerated literal), it is possible to find : a(b) where a is a label and b is an enumeration literal, or : a(b) where a is an entity and b is an architecture. The difference is made by the '@'.

(delays, for example). This file would be produced by some layout extractor. The model could read the file and self-attach the delays to the appropriate statements. This is doable at the expense of parsing the whole file in every place where we need a delay:

```
impure function FIND_DELAY(INSTANCE : STRING) return TIME is
begin
-- here, depending on the syntax of the file:
-- open the file
-- read the file until the INSTANCE is recognized
-- read the associated delay
-- return the delay
end FIND_DELAY;

...

S <= VALUE after  FIND_DELAY(S'INSTANCE_NAME);
```

This inefficiency can be repaired by:

- use of implementation-dependent binding of the delays
- use of implementation-dependent packages (typically one that contains the magic function), or
- sophisticated hash or sort methods programmed in VHDL, a subject that is obviously beyond the scope of this book.

Introduction
New Simulation Mechanisms
New Structuring Mechanisms
New Interfacing Mechanisms
New Predefined Operators, Functions & Attributes
=> ***Slight Enhancements***
Language Simplifications
Clarifications
Annex

18. INERTIAL SIGNAL ASSIGNMENT STATEMENT

LRM REFERENCES: 8.4

18.1. BACKGROUND

VHDL'87 proposes two different models for signal propagation delay. The transport model corresponds to a pure propagation delay where an input waveform is propagated with no distortion: any pulse, however small, is transmitted. This mode is explicitly specified by appending the reserved word **transport** to the signal assignment symbol (<=).

The default signal propagation is the inertial delay mode. In this mode, a pulse smaller than the inertial delay is rejected. As a consequence, all spikes are filtered out. This mode is very useful in digital circuit modeling, where pulses arriving at gates do not propagate if they are smaller than a certain width. However, two flaws can be noted in this handling of propagation delays:

- First, the inertial mode is only implicit, and there is no way to explicitly specify it, which could be useful for documentation purposes.
- Second, there is no way to specify an inertial delay that is different from the propagation delay, because VHDL'87 only considers the earliest transaction propagation delay as the inertial delay. In the following example:

```
S <= A after 5 ns, B after 100 ns;
```

the earliest transaction will occur in 5 ns, this value is taken as the inertial delay (i.e., any pulse smaller than 5 ns is rejected and will not appear on S).

While maintaining upward compatibility with VHDL'87, VHDL'92 adds new constructs and reserved words to overcome these two deficiencies.

18.2. DESCRIPTION

The LRM offers a new syntax to the signal assignment statement:

```
signal_assignment_statement ::=    [LRM 8.4]
[ label : ] target <= [ transport | [ reject time_expression ] inertial ] waveform ;

waveform ::= waveform_element { , waveform_element } |  unaffected
```

Figure 18.1 summarizes the different signal assignment possibilities. It is now possible to use the reserved word **inertial** to point out the inertiality of a statement (see SIGNAL1). The reserved word **reject** (see SIGNAL2) may be used to specify a pulse rejection limit different from the time expression associated with the first waveform element. The compiler will generate an error if the value given in the rejection clause is greater than that of the first waveform element.

It is important to note that this new feature does not add any new functionality to VHDL. The new concepts can be achieved in VHDL'87, but require an extra signal and two concurrent signal assignment statements. One of these is an inertial assignment to an intermediate signal with a propagation delay equal to the delay specified by the initial rejection clause. The effect of this first assignment is to filter the input waveform according to the specified inertial delay (the original pulse rejection limit). The second is a pure transport assignment, with the propagation delay set to the difference between the initial propagation and the rejection clause delay.

The VHDL'92 statement

```
SIGNAL1 <= reject 3 ns inertial REF after 10 ns;
```

is equivalent to the following two *concurrent* statements involving an extra signal (here, Signal2):

```
SIGNAL2 <= REF after 3 ns;                  -- inertial delay pulse filtering
SIGNAL1 <= transport SIGNAL2 after 7 ns;    -- pure propagation delay (10 ns - 3 ns)
```

Therefore, this new feature allows a more concise code.

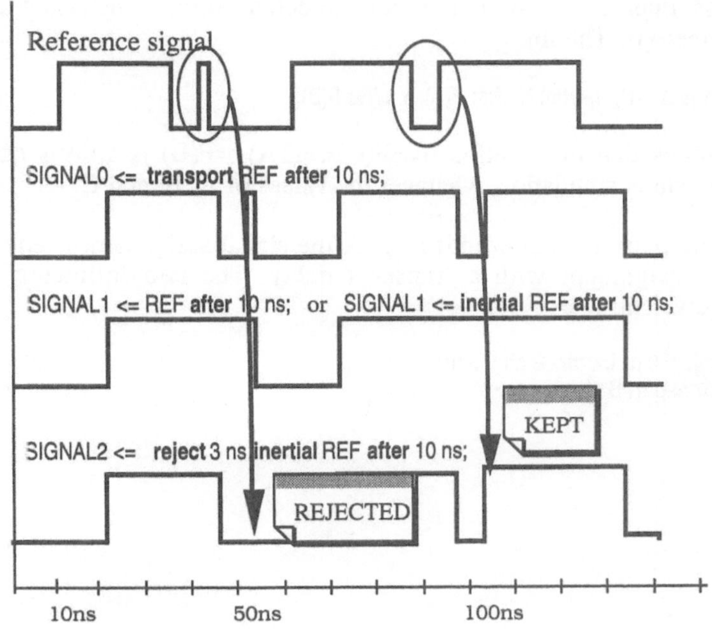

Fig 18.1 Different Signal Assignment Capabilities

18.3. USING THE FEATURE

Even if using such a statement seems straightforward, some comments are necessary:

- It seems to be a good practice, from a documentation point of view, to use the reserved word **inertial** explicitly when some inertial signal assignments are among transport signal assignments. If the entire design uses only inertial signal assignments, the implicit form seems better.

- Simulation errors may occur when using dynamic expressions within a signal assignment. This was true in VHDL'87 when writing

```
S <= X after F(A), Y after F(B);
```

where F(A) and F(B) must always be positive and F(A) must always be less than F(B). With the new rejected form, constraints are more numerous. The line

```
S <= reject F(C) inertial X after F(A), Y after F(B);
```

implies that the relation 0<=F(C)<=F(A)<=F(B) is always true during the whole simulation, whatever the values of A, B, and C.

- If the pulse rejection limit is zero, the signal assignment is equivalent to an assignment with a transport delay. The two following lines are equivalent:

```
S <= reject 0 ns inertial B after 5 ns;
S <= transport B after 5 ns;
```

19. DECLARATIVE PART IN GENERATE STATEMENTS

LRM REFERENCES: 9.7

19.1. BACKGROUND

The VHDL'87 Language Reference Manual defines the generate statement with the following syntax:

```
generate_statement ::=   generate_label : generation_scheme generate
                             { concurrent_statement }
                         end generate [ generate_label ] ;
```

The elaboration of the generate statement is described as follows: it "consists of the replacement of the generate statement with zero or more copies of a *block statement* whose statement part consists of the concurrent statements contained within the generate statement." The elaboration description, using blocks statements, shows that the generate statement constitutes a declarative region, even if this construct does not include any declarative part.

Actually, in the case of a generate statement with a **for** generation scheme, each generated block statement (one for each value of the discrete range) includes in its declarative part a single constant declaration with the same simple name as that of the generate parameter. For example, the third iteration (N=3) of the next generate statement

```
C : for N in 1 to 5 generate
        CI : COMP_NAME port map (S(N), T(N));
        S(N) <= not V(N) after 5 ns;
end generate C;
```

is replaced at elaboration time by the following block statement:

```
C : block
        constant N : INTEGER := 3;
begin
        CI : COMP_NAME port map (S(N), T(N));
        S(N) <= not V(N) after 5 ns;
end block C;
```

The constant N, implicitly declared in the generation scheme of the generate statement, is only visible inside the generate statement (or inside the equivalent block statement). This allows us to write two consecutive generate statements with the same generate parameter name:

```
C1 : for N in 1 to 5 generate
        CI : COMP1 port map (...);
end generate C1;
C2 : for N in 1 to 5 generate
        CI : COMP2 port map (...);
end generate C2;
```

Even if this is not specified in the LRM, the generate statement must be handled by compilers as a declarative region.

Because of this assumption, and because the VHDL'87 generate statement does not include a declarative part, some problems appear when writing configuration specifications for component instantiations located immediately within a generate statement. Considering the next design unit,

```
entity ENT is
        port (X : in BIT; Y : out BIT);
end ENT;

architecture ARC of ENT is
        signal S : BIT_VECTOR(1 to 5);
        component COMP
            port (A : in BIT; S : out BIT);
        end component;
```

```
begin
        L0 : S(1) <= X;
        L1 : for N in 1 to 4 generate
              L2 : COMP port map (S(N), S(N+1) );
        end generate;
        L3 : COMP port map (S(5), Y );
end ARC;
```

the only correct way to write a configuration specification for the component instance L2 inside the generate statement L1 is to write a block statement inside this generate statement, to include the component instantiation inside the statement part of the block statement, and to write the configuration specification inside the declarative part of the block statement:

```
architecture ARC of ENT is
        signal S : BIT_VECTOR(1 to 5);
        component COMP
              port (A : in BIT; S : out BIT);
        end component;
        for L3 : COMP use entity WORK.SHIFT(BEHAVIOR); -- configuration of L3
begin
        L0 : S(1) <= X;
        L1 : for N in 1 to 4 generate
              C : block
                    for L2 : COMP use entity WORK.SHIFT(BEHAVIOR); -- configuration of L2
              begin
                    L2 : COMP port map (S(N), S(N+1) );
              end block;
        end generate;
        L3 : COMP port map (S(5), Y );
end ARC;
```

But this workaround is definitely too verbose to be recommended. On the other hand, some VHDL'87 compilers process the configuration specification written in the next example without handling the generate statement as a declarative region:

```
architecture ARC of ENT is
        signal S : BIT_VECTOR(1 to 5);
        component COMP
              port (A : in BIT; S : out BIT);
        end component;
        for all : COMP use entity WORK.SHIFT(BEHAVIOR); -- the configuration specification
```

```
begin
        L0 : S(1) <= X;
        L1 : for N in 1 to 4 generate
             L2 : COMP port map (S(N), S(N+1) );
        end generate;
        L3 : COMP port map (S(5), Y );
end ARC;
```

In this way, the single configuration specification above, using the reserved word **all**, configures all the instances of component COMP found in this architecture, i.e., the instance L2 inside the generate statement (actually, all four generated instances) as well as the instance L3 outside the generate statement.

This is convenient, but is not at all consistent with the generate statement being a declarative region: in the latter case, the labels of the concurrent statements placed inside the generate statement are implicitly declared at the beginning of the declarative region, and thus are unknown outside the generate statement.

The ambiguous definition of the generate statement in the LRM, or more precisely the fact that the generate statement is not explicitly defined as a declarative region, leads to some implementation inconsistencies.

All these issues have been addressed in VHDL'92: the generate statement is now clearly defined as a declarative region, which even includes an optional declarative part.

19.2. DESCRIPTION

Indeed, VHDL'92 allows an optional declarative part on a generate statement. The syntax is the following:

```
generate_statement ::=  generate_label : generation_scheme generate
                              [ generate_declarative_part
                        begin ]
                              { concurrent_statement }
                        end generate [ generate_label ] ;

generate_declarative_part ::= { block_declarative_item }
```

For a generate statement with a **for** generation scheme, the generate loop parameter is implicitly declared by a constant declaration at the beginning of the declarative part.

The semantics of the generate statement is defined in terms of equivalent block statements. For example, the elaboration of the iterative generate statement

```
C : for N in 1 to 5 generate
        signal S : BIT;
begin
        CI : COMP_NAME port map (A(N), S);
        S <= not B(N) after 5 ns;
end generate C;
```

produces for its third iteration the following block statement:

```
C : block
        constant N : INTEGER := 3;
        signal S : BIT;
begin
        CI : COMP_NAME port map (A(N), S);
        S <= not B(N) after 5 ns;
end block C;
```

The VHDL'92 generate statement is clearly defined as a declarative region.

19.3. USING THE FEATURE

As a consequence, the label of any concurrent statement placed inside the generate statement is implicitly declared at the beginning of the declarative part of the generate statement.

Such a label is invisible outside the generate statement: the configuration specification "**for all** : COMP **use entity** WORK.SHIFT(BEHAVIOR);" written in a previous example (architecture ARC of entity ENT) will thus only affect the component instantiation that is at the same level as the generate statement (and is labeled L3), and not those inside the generate statement.

On the other hand, configuration specifications can now be written in the declarative part of the generate statement. The same architecture ARC is thus changed into

```
architecture ARC of ENT is
        signal S : BIT_VECTOR(1 to 5);

        component COMP
            port (A : in BIT; S : out BIT);
        end component;

        for L3: COMP use entity WORK.SHIFT(BEHAVIOR);
```

```
begin
        L0 : S(1) <= X;
        L1 : for N in 1 to 4 generate
            for L2: COMP use entity WORK.SHIFT(BEHAVIOR);
        begin
            L2 : COMP port map (S(N), S(N+1) );
        end generate;
        L3 : COMP port map (S(5), Y );
end ARC;
```

Even if such configuration specifications are more consistent with the concept of declarative regions, they may appear somewhat verbose for the designer, especially in the previous example (if the intent of the designer was to make the same configuration for all instances of component COMP). Nevertheless, any configuration specification related to a component instantiated inside a generate statement can now be written in a clear and unambiguous way.

Of course, the only way to distinguish between the different generated instances (to achieve a possibly different configuration for each instance) is still to write a configuration declaration for the previous design unit. Assuming that any configuration specification has been removed from the previous architecture ARC, the next design unit can then be written:

```
configuration CONF of ENT is
    for ARC
        for L1(1 to 2)  -- the two first instances are using entity SHIFT1
            for L2 : COMP use entity WORK.SHIFT1(BEHAVIOR);
            end for;
        end for;
        for L1(3 to 4)  -- the two other instances are using entity SHIFT2
            for L2 : COMP use entity WORK.SHIFT2(BEHAVIOR);
            end for;
        end for;
        for L3 : COMP use entity WORK.SHIFT3(BEHAVIOR);
        end for;
    end for;
end CONF;
```

This configuration declaration includes a single block configuration ("**for** ARC... **end for**;"), which itself includes two block configurations (with the two block specifications "L1(1 **to** 2)" and "L1(3 **to** 4)") and one component configuration (with the component specification "L3 : COMP").

Any block specification written in a block configuration can indeed be a generate statement label followed by an index specification (like "L1(1 **to** 2)" in the example above). Such specification is not permitted by the syntax of the configuration specification (in VHDL'87 as well as in VHDL'92).

Finally, since the generate declarative part may include any block declarative item, local objects (such as signals) can be declared and used inside the generate statement, as in any block statement.

Implementing a declarative part in the generate statement should not represent a very complex task for the VHDL tool builder. Indeed, most VHDL'87 tools already consider the generate statement as a kind of declarative region. Thus, adding an optional declarative part does not represent an important implementation change.

20. MAPPING EXPRESSIONS TO INPUT PORTS

LRM REFERENCES: 5.2.1.2

20.1. BACKGROUND

VHDL'87 requires actual parts of port associations to be the reserved word **open** or to denote signals. In the case of an **in** port, the **open** association is allowed only if the corresponding formal **in** port declaration includes a default expression, in which case the port has a single driver with a constant value set to the default expression. That constant driver is used during simulation by the kernel to compute the effective value of the **in** port.

As a consequence of the above VHDL'87 rules, it is not possible to associate formal ports of mode **in** with expressions in a port association. In addition to unnecessarily limiting the flexibility of port associations, this results in a useless multiplication of signals to hold constant values.

The solution proposed by VHDL'92 extends the semantics of the open association of formal ports of mode **in**. That is, a formal port of mode **in** can be associated with an expression that will be used as the constant value of the single driver that will be created for the formal port during elaboration. Because the expression is evaluated during the initialization phase to define the initial value of the driver, it must be globally static, i.e., elaboration-time evaluable.

20.2. DESCRIPTION

The semantics of associating formal **in** ports with expressions is equivalent to an unconnected input port (**open**) with this expression as the default value.

The VHDL'92 text

```
component COMP is          -- "is" here, see chapter 29
        port(INPUT1,INPUT2 : in BIT; OUTPUT : out BIT);
end component COMP;     -- "COMP" to repeat the component name, see also chapter 29
...
signal INPUT, OUTPUT : BIT;
...
begin
...
C : COMP port map (INPUT, '0', OUTPUT);
...
```

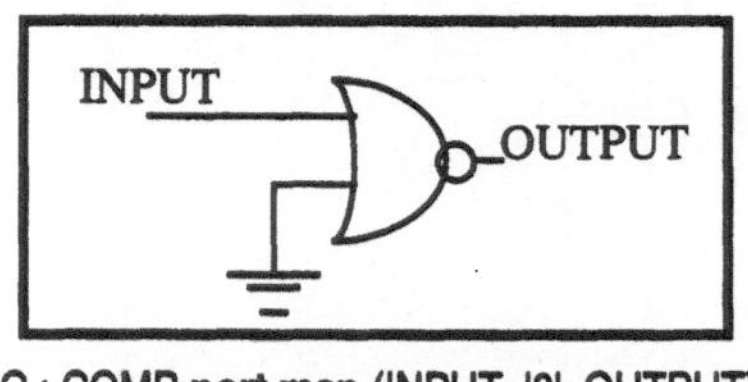

C : COMP **port map** (INPUT, '0', OUTPUT);

Fig 20.1 Mapping Expression to the Input Port

is equivalent to the VHDL'87 text

```
component COMP     -- No "is" here
        port(     INPUT1,INPUT2 : in BIT := '0'; -- default expression necessary for open port
                  OUTPUT : out BIT);
end component; -- No component name here (COMP)
...
signal INPUT, OUTPUT : BIT;
...
begin
...
C1: COMP port map (INPUT, open, OUTPUT);
...
```

In the case where the same component declaration should be instantiated with different expressions for the same input port, it is natural using VHDL'92 to write

```
C2: COMP port map (INPUT,'0', OUTPUT);
C3: COMP port map ('1',INP2, OUTPUT);
C4: COMP port map ('0','1', OUTPUT);
```

Since a formal port declaration can include only one default value, the same functionality is possible in VHDL'87 but requires an extra signal declaration to hold each distinct value used. For the above example, we need a signal to hold the value '1':

```
        signal ONE : BIT :='1';
begin
        ...
        C1: COMP port map (INPUT, open, OUTPUT);
        C2: COMP port map (ONE , INP2, OUTPUT);
        C3: COMP port map (open, ONE , OUTPUT);
        ...
```

The expression associated with the input port must be elaboration-time known. This means that its value must be computable before time zero of the simulation; for example, it can use arbitrarily complex expressions involving (user-defined) function calls, generic values, or constants, but no reference to the value of a signal. This disallows the situation shown in figure 20.2.

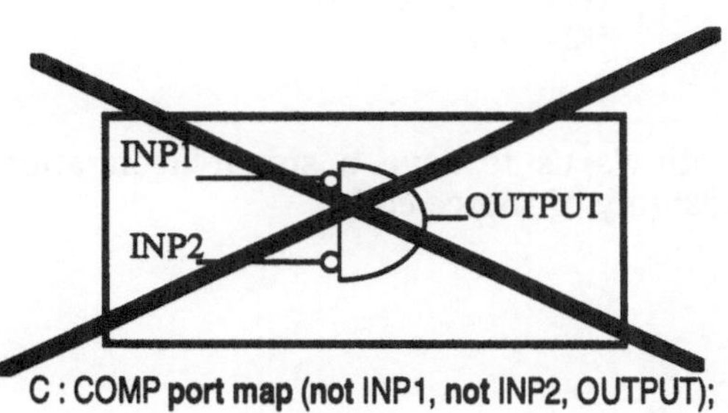

Fig 20.2 Illegal Mapping of Expression to Input Ports

20.3. USING THE FEATURE

Here is a typical example of the use of a JK in place of a divider by two. The basic equation for the JK is

$$Q = J\overline{Q} + \overline{K}Q$$

where the Q (on the left) is the new value of Q computed from the "old" value of Q and from the current value of J and K. This "new" value is synchronously computed at each clock period.

The consequence is that, if J is set to '1' and K to '1', the equation becomes

$$Q = \overline{Q}$$

which may be seen as a synchronous inverter feeding itself back, i.e. a divider by two of the CLOCK:

```
component JK is
        port (J,K,CLOCK : in BIT; Q,NQ : out BIT);
end component;

begin
DIVIDER : JK port map (  J=> '1',
                         K=> '1',
                         CLOCK => INPUT,
                         Q => DIVIDED_OUTPUT,
                         NQ => open);
...
```

This feature allows us to save a signal declaration, which increases efficiency and readability of source code.

21. THE NEW CHARACTER SET

LRM REFERENCES: 14.2

21.1. BACKGROUND

The character set used by VHDL was taken from that defined for Ada in the early 1980s. It was based on a 128-position ISO 646 coded on seven bits.

Ada and VHDL have the characteristics of the languages defined in these earlier days, when some terminals were not even able to handle lower-case characters. In the same spirit, alternate characters (LRM'87 13.10) were defined for terminals lacking the vertical bar, the sharp, or the quotation characters.

The LRM'87 states (LRM'87 13.1) : "The only characters allowed in the text of a VHDL description are the graphic characters and format effectors." This statement is very restrictive, since it includes strings and comments, both of which are part of "the text of a VHDL description."

For VHDL'92 It was decided to extend the character set to 256 positions, therby permitting the handling of identifiers, strings, or comments, including characters like â, ö, ß, å, æ, or ÿ. This might create interesting problems for implementors using lexer generators, which often represent the end of a file or the end of a line by the use of special "magic" characters.

21.2. VHDL'87/VHDL'92

The "old" character set was defined in the package STD.STANDARD in the manner shown in figure 21.1:

```
type CHARACTER is ( -- VHDL 87
NUL,  SOH,  STX,  ETX,  EOT,  ENQ,  ACK,  BEL,
BS,   HT,   LF,   VT,   FF,   CR,   SO,   SI,
DLE,  DC1,  DC2,  DC3,  DC4,  NAK,  SYN,  ETB,
CAN,  EM,   SUB,  ESC,  FSP,  GSP,  RSP,  USP,
' ',  '!',  '"',  '#',  '$',  '%',  '&',  ''',
'(',  ')',  '*',  '+',  ',',  '-',  '.',  '/',
'0',  '1',  '2',  '3',  '4',  '5',  '6',  '7',
'8',  '9',  ':',  ';',  '<',  '=',  '>',  '?',
'@',  'A',  'B',  'C',  'D',  'E',  'F',  'G',
'H',  'I',  'J',  'K',  'L',  'M',  'N',  'O',
'P',  'Q',  'R',  'S',  'T',  'U',  'V',  'W',
'X',  'Y',  'Z',  '[',  '\',  ']',  '^',  '_',
'`',  'a',  'b',  'c',  'd',  'e',  'f',  'g',
'h',  'i',  'j',  'k',  'l',  'm',  'n',  'o',
'p',  'q',  'r',  's',  't',  'u',  'v',  'w',
'x',  'y',  'z',  '{',  '|',  '}',  '~',  DEL);
```

Fig 21.1 The VHDL'87 Character Type

There is very little to say about this definition. To avoid collision with the units of the type TIME, some characters have been renamed from their ISO name (e.g., FS into FSP). The "non-printable" characters hold a name that is an ordinary identifier, directly visible from any design unit.

The "new" character set is now defined as shown in fig. 21.2.

Positions 128 to 255 now contain the accents, tildes, and other national characters.

An important point is that in VHDL'87 it was already possible to define, an EXTENDED_CHARACTER type, on top of the existing CHARACTER. For example, one could write:

```
package FOR_EXTENDED_CHARACTER is
    type EXTENDED_CHARACTER is ( -- VHDL'87 workaround
        NUL,   SOH,   STX,   ETX,   EOT,   ENQ,   ACK,   BEL,
        -- here 128 positions defined like in CHARACTER
        'x', 'y', 'z', '{', '|', '}', '~', DEL,
        C128,  C129,  C130,  C131,  C132,  C133,  C134,  C135,
        -- here positions up to C159
        NBSP,  C161,  C162,  C163,... SHY,...     C255);
end FOR_EXTENDED_CHARACTER;
```

```
type CHARACTER is ( -- VHDL 92
NUL,  SOH,  STX,  ETX,  EOT,  ENQ,  ACK,  BEL,
BS,   HT,   LF,   VT,   FF,   CR,   SO,   SI,
DLE,  DC1,  DC2,  DC3,  DC4,  NAK,  SYN,  ETB,
CAN,  EM,   SUB,  ESC,  FSP,  GSP,  RSP,  USP,
' ',  '!',  '"',  '#',  '$',  '%',  '&',  ''',
'(',  ')',  '*',  '+',  ',',  '-',  '.',  '/',
'0',  '1',  '2',  '3',  '4',  '5',  '6',  '7',
'8',  '9',  ':',  ';',  '<',  '=',  '>',  '?',
'@',  'A',  'B',  'C',  'D',  'E',  'F',  'G',
'H',  'I',  'J',  'K',  'L',  'M',  'N',  'O',
'P',  'Q',  'R',  'S',  'T',  'U',  'V',  'W',
'X',  'Y',  'Z',  '[',  '\',  ']',  '^',  '_',
'`',  'a',  'b',  'c',  'd',  'e',  'f',  'g',
'h',  'i',  'j',  'k',  'l',  'm',  'n',  'o',
'p',  'q',  'r',  's',  't',  'u',  'v',  'w',
'x',  'y',  'z',  '{',  '|',  '}',  '~',  DEL,
C128, C129, C130, C131, C132, C133, C134, C135,
C136, C137, C138, C139, C140, C141, C142, C143,
C144, C145, C146, C147, C148, C149, C150, C151,
C152, C153, C154, C155, C156, C157, C158, C159,
' ',  '¡',  '¢',  '£',  '¤',  '¥',  '¦',  '§',
'¨',  '©',  'ª',  '«',  '¬',  '-',  '®',  '¯',
'°',  '±',  '²',  '³',  '´',  'µ',  '¶',  '•',
'¸',  '¹',  'º',  '»',  '¼',  '½',  '¾',  '¿',
'À',  'Á',  'Â',  'Ã',  'Ä',  'Å',  'Æ',  'Ç',
'È',  'É',  'Ê',  'Ë',  'Ì',  'Í',  'Î',  'Ï',
'•',  'Ñ',  'Ò',  'Ó',  'Ô',  'Õ',  'Ö',  '5',
'Ø',  'Ù',  'Ú',  'Û',  'Ü',  '•',  '•',  'ß',
'à',  'á',  'â',  'ã',  'ä',  'å',  'æ',  'ç',
'è',  'é',  'ê',  'ë',  'ì',  'í',  'î',  'ï',
'•',  'ñ',  'ò',  'ó',  'ô',  'õ',  'ö',  'Π',
'ø',  'ù',  'ú',  'û',  'ü',  '•',  '•',  'ÿ');
```

Fig 21.2 The VHDL'92 Character Type[16]

The differences between this workaround and the new capability are as follows:

- This type is not named CHARACTER and is not in the package STANDARD: one must to use the package and rename CHARACTER as EXTENDED_CHARACTER.
- Most of the new positions cannot be defined with their graphical equivalent (e.g., C255 instead of 'ÿ'). In the same way, character literal

[16] Except the first one, characters represented by a '•' have no representation in the font used here (Monaco 9). The graphic representation of characters is not part of the standard.

cannot be used in a string: "château" has to be written: "ch" & C226 & "teau."[17]

- More importantly, TEXTIO does not work on this type. One must either convert EXTENDED_CHARACTER to CHARACTER whenever possible (using an equivalence table, for example), or rely on an implementation-dependent EXTENDED_TEXTIO.

21.3. USING THE FEATURE

The main aspect of this new feature is that it is **not upward compatible**.

Some attributes of the type CHARACTER have changed, e.g., HIGH, RANGE, RIGHT, and some cases that were error conditions are now valid (see figure 21.3).

Some new attributes have a special action with character literals: 'VALUE and 'IMAGE convert the graphic aspect of the characters from and to their value in terms of elements of the type: CHARACTER'VALUE(" 'A' ") is 'A'.

CHARACTER'IMAGE('A') is the string "'A'" (with the simple quotes). Surprisingly, the converted string includes the simple quotes of the character literal. But, deeper consideration shows that, it is symmetric to the action on identifiers:

```
type FOO is (ONE,'2');
```

FOO'IMAGE(ONE) is the string "ONE" as it appears in the type. FOO'IMAGE('2') is the string "'2'" as it appears in the type. Due to the extension of the type, some features of VHDL requiring a full coverage of the type will produce errors in places that were legal before:

```
case A_CHARACTER is
        when NUL to '@'   => null;
        when 'A' to 'Z'   => DO_SOMETHING;
        when '[' to DEL   => DO_SOMETHING_ELSE;
end case;
```

This is legal VHDL'87 (the type is fully covered) and illegal VHDL'92 (the **others** clause is missing).

Similar examples could be found with the selected assignment statement or with aggregates of arrays indexed by the type CHARACTER.

---

[17] However a nasty trick often used in Ada is to disable some checks of the compiler during the compilation of the unit containing such strings.

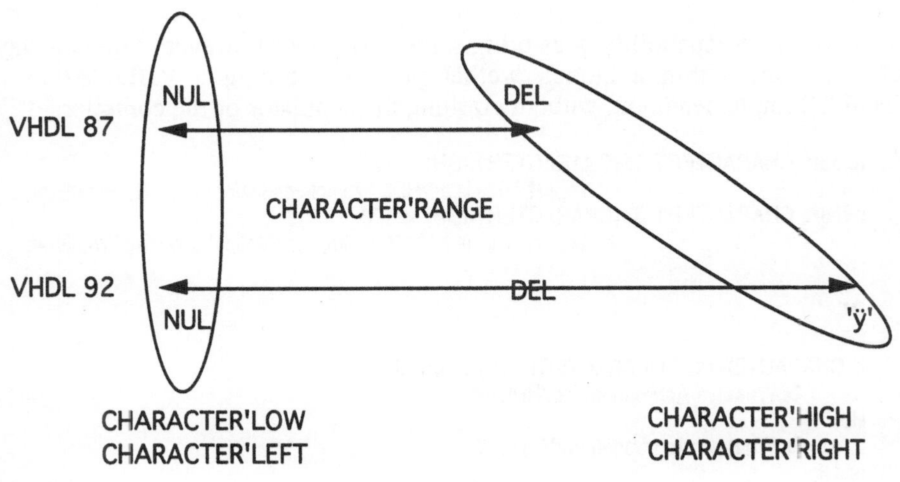

New Attributes

ASCENDING : CHARACTER'ASCENDING = TRUE
IMAGE : see text
VALUE : see text

other VHDL 87 Attributes

POS : same behavior
VAL : CHARACTER'VAL(X) now valid if 127<X<256
SUCC : CHARACTER'SUCC(DEL) now valid
PRED : same behavior
LEFTOF : same behavior
RIGHTOF : CHARACTER'RIGHTOF(DEL) now valid
REVERSE_RANGE : see RANGE in figure

Fig 21.3 VHDL'87 Versus VHDL'92 Attributes

21.4. TRICK

This non-compatibility does bring some good news however: here is a way to test, from within a model, wether you are running a VHDL'87 or a VHDL'92 implementation, without crashing the simulator or the compiler:

```
assert CHARACTER'POS(CHARACTER'HIGH) > 127
                        report "This is a VHDL 87 implementation" severity WhatEver;
assert CHARACTER'POS(CHARACTER'HIGH) < 128
                        report "This is a VHDL 92 implementation" severity WhatEver;
```

or

```
if CHARACTER'POS(CHARACTER'HIGH) > 127 then
    DoWhateverAppropriateForVHDL'92;
else
    DoWhateverAppropriateForVHDL'87;
end if;
```

22. IDENTIFIER GENERALIZATION

LRM REFERENCES: 13.3.2

22.1. BACKGROUND

One well-known problem in using VHDL'87 to model existing hardware devices is that, sometimes, their names and ports are not legal VHDL identifiers. These names (74LS238, BUFFER, A#12, etc.) begin with a digit, conflict with VHDL reserved words, or contain special characters. The designer must use other names, and this does not help to clarify the interface with such a device.

Up to now, there have been many restrictions on the syntax of VHDL identifiers. Nothing is more frustrating for a designer than to have to change the name of a pin because it is in conflict with a reserved word or begins with a digit. In VHDL'92, the notion of "extended identifiers" solves such a problem. We will refer to the previous forms of identifiers, which of course are always legal, as the "short identifiers" LRM [13.3.1].

22.2. DESCRIPTION

An extended identifier has the following characteristics:

- It is delimited with backslashes:
  \IDENT\ is an extended identifier.

- It may include any graphic character but no format effectors (CR, LF, FF, etc.). Nevertheless, the space is allowed:
  \A and B\, \&rty#Ë\ are legal extended identifiers.

- It may be identical to a reserved word:
  \BUFFER\ is a legal extended identifier.

- It may begin with a digit:
  \5600B\ is a correct extended identifier.

- It may include two (or more) adjacent underscores:
  \BUFFER_____1\ is a legal extended identifier (to be avoided).

- It is case sensitive:
  \ABC\ is different from \ABc\. Both are different from \abc\.

- It is always different from any short identifier:
  \ABC\ is different from ABC and from abc.

- A pair of adjacent backslashes represents a backslash in the name of the identifier:
  \AB\\C\ represents the name AB\C (four characters).

22.3. USING THE FEATURE

The use of these identifier is the same as for short ones. The benefit of this improvement is clear for the designer. It is now possible to give the name that seems the most natural to an item of the model without the previous restrictions. As shown in the following example, the readability of the source code will obviously be improved.

```
        ...
component \74S405\ is -- "is" is now possible in VHDL'92 (cf. chapter 29)
        port (\1N2\: in BIT;
        ....
end component;
        ....
begin
        C1: \74S405\ port map ...
```

A question could be asked:

- How can we use the predefined attributes 'SIMPLE_NAME, 'IMAGE, or 'VALUE(see chapters 16 & 17) on such an identifier? In both cases, the backslashes are part of the string that is read or written.

Trick: When using a VHDL'92 parser on VHDL'87 source code, some errors may occur due to identifiers that are new reserved words in VHDL'92 version. The reserved words **group**, **impure**, **inertial**, **literal**, **postponed**, **pure**, **reject**, **rol**, **ror**, **shared**, **sla**, **sll**, **sra**, **srl**, **unaffected**, and **xnor** are now prohibited as identifiers. Since such identifier names may have a given "semantics" for the designer, a good practice is to change them to extended names. In this case, XNOR will become \XNOR\, will pass the compilation and will keep the same power of documentation as previously expected. It is nevertheless important to notice that this new identifier is now case sensitive: \XNOR\ is different from \xnor\ and from \Xnor\.

23. ALIAS GENERALIZATION

LRM REFERENCES: 4.3.3

23.1. BACKGROUND

The use clause provides a general way to import declarations, but it is impossible in VHDL'87 to export declarations that have been imported with a use clause. Such a mechanism of exportation is useful when defining new interfaces for already existing packages. One way to export imported declarations is to use alias declarations, but the VHDL'87 aliases are too restricted. Indeed, the alias declaration makes it possible to give an alternate name for an *object* (i.e., a signal, a variable, or a constant), but not for types or for subprograms.

23.2. DESCRIPTION

VHDL'92 generalizes the alias declaration to any declared name in order to be able to export any imported declaration. The syntax of the alias declaration is generalized as follows:

alias_declaration ::= **alias** alias_designator [: subtype_indication] **is** general_name ;

alias_designator ::= identifier | operator_symbol | character_literal

general_name ::= name [signature]

signature ::= **[**[18] [type_mark { , type_mark }] [**return** type_mark] **]**

The signature introduces in the syntax the notions of parameter and result type profile that are already present in the semantics of VHDL'87. It is used to eliminate ambiguity in overloaded subprogram and enumeration literal names.

A signature is allowed in the alias declaration only when the name is indeed a subprogram designator or an enumeration literal. It is required if the subprogram or enumeration literal is overloaded: the parameter and result type profile specified by the signature must then match the parameter and result type profile of a subprogram or enumeration literal with the same simple name (matching in the sense of the base types, as in VHDL'87).

Only the base types of the arguments, and all of them, are used in the alias. In particular, if a formal argument has a default value, this does not mean that the corresponding type mark in the alias can be omitted.

The optional subtype indication is never required and is only allowed when the name denotes an object (constant, variable, or signal).

If this subtype indication is present, the alias declaration has the same semantics as the VHDL'87 alias declaration. Otherwise, the alias declaration declares a new designator for an already declared item. Furthermore, if the aliased item is a type declaration, then any enumeration literals, predefined operators, and physical units associated with the type are immediately given alias (without change in their designator) after the alias declaration.

Finally, an alias designator can be a character literal only when the general name denotes a character literal, and can be an operator symbol only when the general name denotes an operator.

23.3. USING THE FEATURE

Following are typical examples of alias declarations. First, we give an object alias declaration (identical to the alias declaration in VHDL'87):

---

[18] Do not confuse the '[' and ']' (bolded) which are lexical elements of VHDL'92, with the '[' and ']' which indicate an option in the syntax itself.

```
package MICROPROCESSOR is
        signal INSTRUCTION_REGISTER : BIT_VECTOR(0 to 15);
        alias OP_CODE : BIT_VECTOR(1 to 5) is INSTRUCTION_REGISTER (0 to 4) ;
        -- alias of a signal: same as VHDL'87 alias declaration
end;
```

In VHDL'92, the subtype indication in the alias declaration is optional, even when the aliased name denotes an object:

```
architecture STRUCTURE of DATAPATH is
        signal DATABUS : BIT_VECTOR(31 downto 0);
        alias SIGN is DATABUS (31);
        -- no subtype indication: SIGN is of the subtype of DATABUS (31), i.e., BIT
begin ...
end architecture STRUCTURE ;
```

Aliases of subprograms, even *overloaded* subprograms, can be declared:

```
package ARITH is
        type XBIT is ('X', '0', '1') ;
        function "or"( A, B : XBIT ) return XBIT ;
        alias "+" is "or" [XBIT, XBIT return XBIT] ;
end package ARITH ;
```

Finally, the three packages below illustrate the use of these generalized alias declarations to export imported declarations:

```
package COUNT is  -- original package
        type DIGIT is ('0', '1', '2', '3', '4', '5', '6', '7', '8', '9');
        function VALUE(A : DIGIT) return NATURAL;
        function VALUE(A : CHARACTER) return NATURAL;
end package COUNT;

package IMPEXP is  -- package exporting declarations imported from package COUNT
        alias NUMBER is WORK.COUNT.DIGIT;
        -- alias of a type: enumeration literals of the type are implicitly given alias here:
        -- alias  '0' is WORK.COUNT.'0' [return WORK.COUNT.DIGIT];
        -- alias  '1' is WORK.COUNT.'1' [return WORK.COUNT.DIGIT]; -- and so on, up to:
        -- alias  '9' is WORK.COUNT.'9' [return WORK.COUNT.DIGIT];
        alias VAL is WORK.COUNT.VALUE[WORK.COUNT.DIGIT return NATURAL];
        -- alias of a function (signature is used)
end package IMPEXP;

use WORK.IMPEXP.all;
package USING is
        procedure PRINT( A : NUMBER );
        constant ZERO : NATURAL := VAL( NUMBER 'LEFT );
end package USING;
```

24. ACCESS TO PREDEFINED OPERATORS

LRM REFERENCES: 4.3.3 (ALIAS)

24.1. BACKGROUND

It is often desirable to overload predefined operators. For example, one might want to transform the arithmetic "+" into an addition modulo something, or to make the "*" a multiplication with a controlled precision.

Overloading a predefined operator is a well-known facility of VHDL. In VHDL'87, the definition of an overloaded operator makes the predefined operator unreachable in the region where the new operator is declared. No mechanism such as the dotted notation is available to select the predefined operator.

Two main problems may occur in this respect:

- One might want to use the predefined and the overloaded operator in the same region.
- Inside the very definition of the new operator, it might be convenient to use the old one.

The following text is incorrect VHDL because the "=" used inside the function will be the "=" currently defined, which leads to an infinite recursion.

```
type MVL is ('0','1','X');
-- implicit declaration of :
-- function"="(anonymous,anonymous : MVL) return BOOLEAN; -- among other functions

function"="(LEFT,RIGHT : MVL)    return BOOLEAN is
-- the only "="(MVL,MVL) visible is now this very function
begin
  if LEFT='X' or RIGHT='X' then return FALSE;
  else return LEFT = RIGHT;
  end if;
end"=";
```

One workaround is to use a truth table:

```
function "=" (LEFT,RIGHT : MVL) return BOOLEAN is
        type MVL_MATRIX is array (MVL,MVL) of BOOLEAN;
        constant TT : MVL_MATRIX :=      ((TRUE,FALSE,FALSE),
                                          (FALSE,TRUE,FALSE),
                                          (FALSE,FALSE,FALSE));
begin
        return TT(LEFT,RIGHT);
end "=";
```

This method works, but there is no general rule about the compared efficiency of the algorithm versus the truth table. In any case, there are examples where it is obviously better to use the algorithm (nearly regular results) or to use the table (very random results). It may be desirable, or even necessary, to have access to predefined operators, even when they are hidden.

Another workaround was to use an intermediate function:

```
type MVL is ('0', '1', 'X');

function EQUAL (LEFT, RIGHT : MVL) return BOOLEAN is
begin
        return LEFT = RIGHT;
end EQUAL;

function "=" (LEFT, RIGHT : MVL) return BOOLEAN is
begin
        if EQUAL(LEFT, 'X') or EQUAL(RIGHT, 'X') then
            return FALSE;
        else
            return EQUAL(LEFT, RIGHT);
        end if;
end "=";
```

Note that this assumes that the function "=" is not declared before the function EQUAL; in particular, it cannot be exported in a package declaration.

A final workaround in VHDL'87 is to arrange the various declarations in different declarative regions in order to distinguish the operators by their names. This imply making *ad-hoc* declarative regions. For example,

```
package ONE is
        type MVL is ('0','1','X');
        -- function "=" (LEFT, RIGHT : MVL) return BOOLEAN; -- implicit among other functions
end ONE;

package TWO is
        function "=" (LEFT, RIGHT : WORK.ONE.MVL) return BOOLEAN;
end TWO;

-- Now WORK.ONE."=" and WORK.TWO."=" can be distinguished.
-- Note that WORK.ONE."=" can be used in the body of WORK.TWO."=".
```

24.2. DESCRIPTION

In VHDL'87 only objects (constants, variables, signals) could be given aliases. The alias mechanism has been extended to solve this problem: it is now possible to give an alias to a predefined operator before overloading it. Coming back to our example "=", where EQUAL is the overloaded operator:

```
type MVL is ('0','1','X');
alias EQUAL is "="[MVL,MVL return BOOLEAN];

function "=" (LEFT,RIGHT : MVL) return BOOLEAN is
begin
        if EQUAL(LEFT,'X') or EQUAL(RIGHT,'X') then
            return FALSE;
        else
            return EQUAL(LEFT,RIGHT);
        end if;
end "=";
```

Now this "=" can be declared in a package declaration, and we do not have the burden of an intermediate function.

25. EXTENSION OF BIT STRING LITERALS

LRM REFERENCES: 13.7

25.1. BACKGROUND

The bit string literal as defined in VHDL'87 defines literals of type BIT_VECTOR and can be expressed in binary (B"0101", or "0101"), in octal (O"12") or in hexadecimal (X"1A2"); the problem was that this very convenient notation was not available for other value systems, such as the 9 value system of the 1164 standard.

25.2. DESCRIPTION

In VHDL'92, bit string literals are defined in terms of lexical equivalence with a string literal. Since string literals are potentially compatible with any one-dimensional array of a character type (as long as the elements inside the string conform to the declaration), the binary, octal and hexadecimal notations become available for all value systems.

```
B"1001_1111" is lexically equivalent to "10011111";
O"123" is lexically equivalent to "001010011";
X"A2B" is lexically equivalent to "101000101011";
```

Each example above can be assigned to a signal of type BIT_VECTOR, or of vectors of a multi-valued type like ('X','0','1','Z'), or of STD_LOGIC (STD_LOGIC_1164 package).

Of course, one should not try to introduce inappropriate digits within the string. X"2ZC" is an error.

26. CONCURRENT SIGNAL ASSIGNMENT

LRM REFERENCES: 9.5

26.1. BACKGROUND

The two forms of signal assignment, conditional and selected, are very concise writings of powerful processes. Nevertheless, some aspects of their VHDL'87 syntax can be very confusing.

The designer often wants to make some assignments if some conditions are true and to preserve the present signal value in other cases. Using the conditional form, for example (although the same problem can occur with the selected form), the **else** clause seems convenient, but the result is different from the expected one.

In the following example, we assume that S and CK are previously defined signals of type BIT.

```
S <= '1' after 10 ns when CK='1' else S;
```

This simple statement can be interpreted as follows: signal S receives the value '1' after a delay of 10 ns if the value of the expression CK='1' is true; if not, there is no change in S. This second part of the interpretation is completely wrong.

If the condition (CK='1') is false:

- A transaction is performed on S. The attribute S'TRANSACTION switches.
- The driver of S is erased of its previous values. The signal S itself that is not assigned to S, but rather its current value. All waveforms previously stored in the S driver are lost.

This can lead to logical errors in the VHDL design. What is really needed here is a way to express the fact that, under certain condition, "no change in S" will occur.

The only way to properly solve the "no change" problem in VHDL'87 is to write an explicit process and a sequential signal assignment.[19] The previous example becomes:

```
process
begin
        if CK='1' then
                S <= '1' after 10 ns;
        end if;
        wait on CK;
end process;
```

Although this code takes longer to write, it does not generate any transaction on S (does not modify the driver of S) if the condition (CK='1') is false.

The selected signal assignment can also be modeled in the same way using a select statement.

26.2. DESCRIPTION

VHDL'92 proposes a new reserved word **unaffected** to express the fact that that nothing changes. As shown in figure 26.1, this new construct does not perform any transaction on S when condition COND1 is false. The driver of S remains unchanged.

---

[19] A work-around can be to write :
S <= Value when Condition else Anything after (TIME'HIGH-NOW);

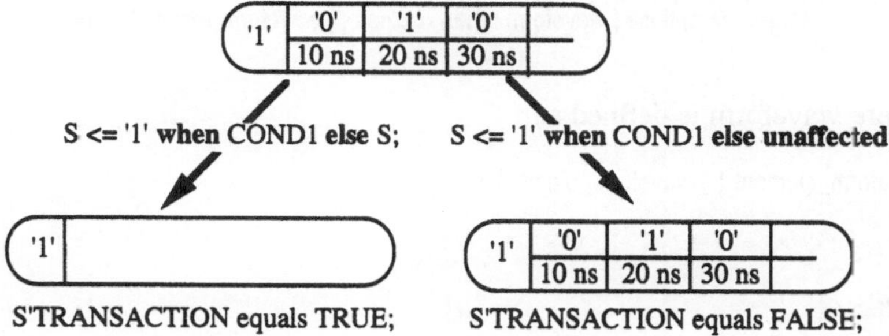

Fig 26.1 Use of the Reserved Word unaffected

Of course, this reserved word can also be used in concurrent selected assignments with the same result. The lines

```
with EXP select
        S <=    '1' after 10 ns when ADD,
                '0' after 10 ns when SUB,
                S when others;
```

have to be replaced by

```
with EXP select
        S <=    '1' after 10 ns when ADD,
                '0' after 10 ns when SUB,
                unaffected when others;
```

Furthermore, in order to allow a simpler, but less explicit, writing of such conditional concurrent statements, the **else** branch is now optional. The two following lines are equivalent:

```
S <= '1' when COND1 else unaffected;
S <= '1' when COND1;
```

The exact syntax of the concurrent signal assignment in VHDL'92 is

```
[ label : ]    [ postponed ]
               target  <= options { waveform when condition else } waveform [ when condition ] ;
```

or

```
[ label : ]   [ postponed ] with expression select
              target  <= options { waveform when choices , } waveform when choices ;
```

where waveform is defined as

```
waveform_element { , waveform_element }
```

or

```
unaffected
```

Note: The notion of postponed concurrent signal assignment is discussed in chapter 3.

26.3. USING THE FEATURE

A simple latch can now be written in a dataflow style:

```
OUTPUT<= INPUT when CLOCK='1' else unaffected;
```

[BER] explains how to use the events of a dedicated signal to avoid the inefficient repetition of such «CLOCK='1'» in many places. It is now simpler to implement:

```
CLOCK_RISES<= not CLOCK_RISES when CLOCK='1' else unaffected;
•••
OUTPUT <= INPUT when CLOCK_RISES else unaffected;
```

26.4. PERFORMANCE IMPLICATION

Since this new writing spares a transaction and the erasing of the driver, the execution of the statement

[BER] VHDL Designer's Reference, Kluwer Academic Publishers 92, Bergé Fonkoua Maginot Rouillard

```
A <= B when COND;
```

is faster than the old one

```
A <= B when COND else A;
```

when the condition COND is false.

27. REPORT STATEMENT

LRM REFERENCES: 8.3

27.1. BACKGROUND

When the flow of control within a process or a subprogram, is being traced, it is sometimes useful to report some information to the designer. VHDL'87 proposes the assertion statement to do so:

```
assert FALSE report "I was here!" severity NOTE;
```

This statement is very cumbersome due to the following facts:

- The condition is mandatory within an assertion statement, and therefore, FALSE has to be written.
- The default value of the severity level is ERROR, which usually stops the simulation. This behavior is not convenient in our case, where the goal is to trace the control flow and continue. The severity level NOTE is more suitable.

27.2. DESCRIPTION

VHDL'92 offers a new sequential statement called "report statement." This statement does not add any functionality to the language, but does allow a shorter version of some forms of the sequential assertion statement.

The syntax is

```
[ label : ] report type_STRING_expression [ severity severity_level] ;
```

As with all other sequential statements, the report statement has an optional label in VHDL'92 (see chapter 35).

The whole semantics of this statement can be described in terms of equivalent sequential assertion statements. The default value of the severity level is NOTE. The two following lines are equivalent:

```
report "Hello";
assert FALSE report "Hello" severity NOTE;
```

as are these:

```
report "Stop" severity ERROR;
assert FALSE report "Stop" severity ERROR;
```

27.3. USING THE FEATURE

As for the assert statement, the new predefined operators 'IMAGE (see chapter 16), 'PATH_NAME, and 'SIMPLE_NAME (see chapter 17) offer new possibilities for messages writing. The value of these attributes is of type STRING.

If we suppose N to be of type INTEGER, the following example will display the number of each iteration of the loop.

```
...
for I in 1 to N loop
...
report "Tracing number "&INTEGER'IMAGE(I);
...
end loop;
...
```

It is also possible to report some information related to the hierarchy:

```
report "I am within "&MY_LABEL'PATH_NAME;
```

28. CONCATENATION OPERATOR

LRM REFERENCES: 7.2.4

28.1. BACKGROUND

There is a very frustrating trap in VHDL'87 when concatenating range arrays: the left bound of the result of the concatenation is the left bound of the left operand. This may lead to constraint error situations.

In the following example, a very simple operation is performed that consists in switching the two bytes of a 16-bit-length word. Due to the value of the left bound of the left operand, the left (and upper) bound of the result is 7. Because of the length of the result (16), the lower bound of the result is deduced to be -8, which is outside the range of the index of the result type (NATURAL for BIT_VECTOR) and causes a constraint error during run-time:

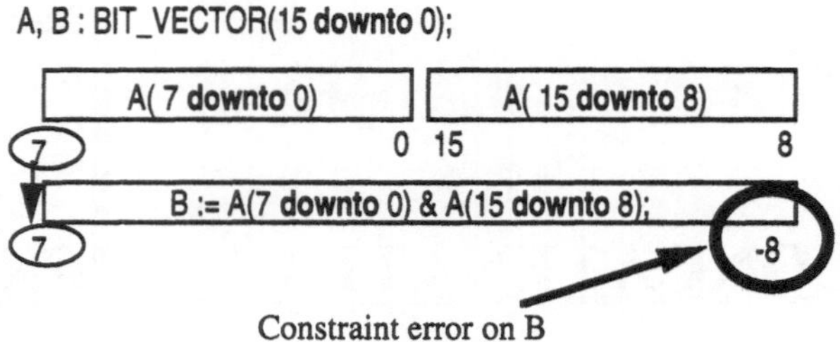

28.2. DESCRIPTION

The new strategy for deducing the bounds of the result array can be summarized by figure 28.1.

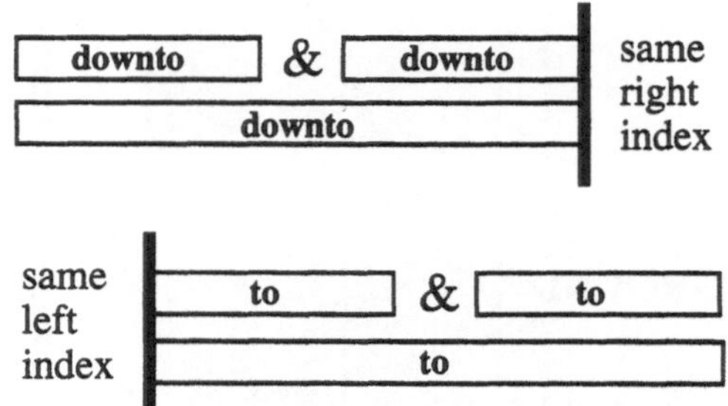

Fig 28.1 Principle of Concatenation in VHDL'92

For ascending arrays, as in VHDL'87, the left bound of the result is the left bound of the left operand. In a symmetrical way, for descending arrays, the right bound of the right operand will be the right bound of the result.

Note: A null array has no effect on the result of such an operation, and in this case the result of the concatenation is the other operand (with its own bounds).

29. BRACKETING: SYNTAX CONSISTENCY

LRM REFERENCES: APP. A

29.1. BACKGROUND

Some VHDL language constructs have a bracket structure delimited by opening and closing markers composed of reserved words. For example, the **if** statement is delimited by the reserve words **if** and **end if**.

```
if -- opening marker
      CONDITION then
      ...--Statements
end if; -- closing marker
```

Such bracketing structures increase the readability of VHDL texts containing nested sets of VHDL constructs. In VHDL'87, the bracketing mechanism is awkward because some constructs (e.g., **if** statement, **component** declaration, **case** statement,etc.) repeat the opening marker after the **end** key word of their closing marker, and others (e.g., **entity** declaration, subprogram declaration,etc.) do not. Furthermore, when the opening marker includes an optional simple name, certain constructs allow it to be repeated and others do not. These inconsistencies can be quite confusing for the beginner, who must remember all the special cases of the bracketing syntax of VHDL.

29.2. DESCRIPTION

VHDL'92 addresses this problem by providing a uniform rule for the bracketing constructs of VHDL, which can be summarized as follows:

- The closing markers are allowed to optionally repeat the reserved word introducing the opening marker.
- If the opening marker includes an identifier, the closing marker is allowed to optionally repeat it.

To meet the design objective of upward compatibility, the modifications allowed are only optional. So the following bracket constructs of VHDL'87 are still valid in VHDL'92 :

component COMP .... **end component**; or : **entity** ENT **is** ... **end** ENT;

29.2.1. Design Unit Declaration Syntax

The syntax for any of the five design units (**entity, architecture, package** declaration, **package** body, and **configuration**) is now modified by allowing the closing marker to *optionally* include a reserved word denoting the class of the design unit. When present, this reserved word must stand after the reserved word **end** and before the optional identifier repeating the name of the design unit. As in VHDL'87, the name of the design unit is optionally added (and must, in this case, match the exact name of the design unit).

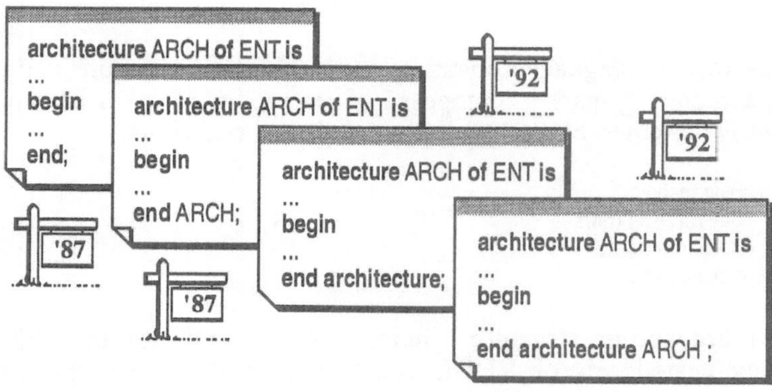

Fig 29.1 Examples of Architecture Body Syntax

The exact syntax of declaration of the five different kinds of design units (respectively entity, architecture body, package, package body and configuration) are:

```
entity identifier is
        entity_header
        entity_declarative_part
[ begin
        entity_statement_part ]
end [ entity ] [ entity_simple_name ] ;

architecture identifier of entity_name is
        architecture_declarative_part
begin
        architecture_statement_part
end [ architecture ] [ architecture_simple_name ] ;

package identifier is
        package_declarative_part
end [ package ] [ package_simple_name ] ;

package body package_simple_name is
        package_body_declarative_part
end [ package body ] [ package_simple_name ] ;

configuration identifier of entity_name is
        configuration_declarative_part
        block_configuration
end [ configuration ] [ configuration_simple_name ] ;
```

29.2.2. Component Declaration Syntax

In VHDL'87, component declaration syntax is commonly very close to that of entity declaration except for the reserved word **is**, not allowed for component declaration. VHDL'92 allow this reserved word to be *optionally* inserted ; likewise the name of the component can be repeated after the final **end component**.

The syntax given in the LRM for component declaration is:

```
component identifier [ is ]
        [ local_generic_clause ]
        [ local_port_clause ]
end component [ component_simple_name ] ;
```

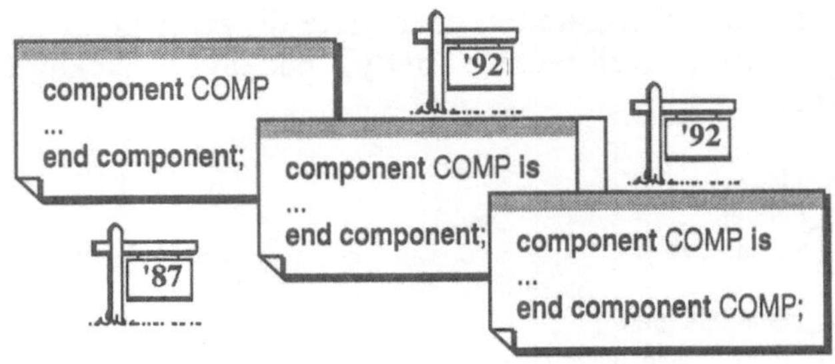

Fig 29.2 Examples of Component Declaration Syntax

29.2.3. Block and Process Declaration Syntax

The reserved word **is** can be *optionally* inserted after **block** or **process** to make these structures consistent with design unit syntax.

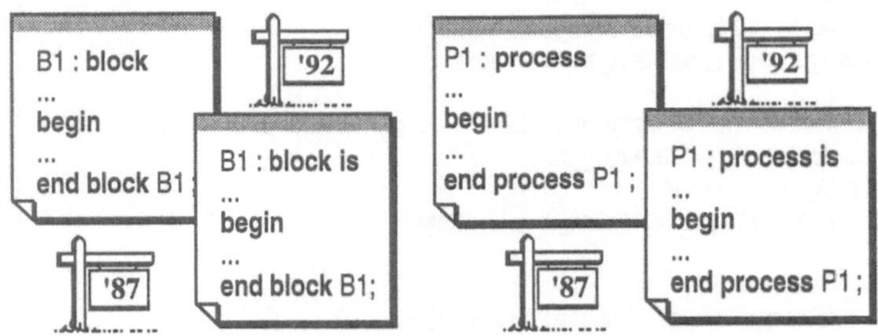

Fig 29.3 Examples of Block and Process Syntax

The general syntax of respectively block and process statement declarations is given bellow:

*block*_label : **block** [(*guard*_expression)] [**is**]
        block_header
        block_declarative_part
begin
        block_statement_part
end block [*block*_label] ;

```
[ process_label : ] [ postponed ] process [ ( sensitivity_list ) ] [ is ]
        process_declarative_part
begin
        process_statement_part
end [ postponed ] process [ process_label ] ;
```

29.2.4. Record Type Declaration Syntax

It is now possible to repeat the name of the record type at the **end** of the type declaration. This can be useful to increase the readability of parts of the source code with a large record declaration (many subelements).

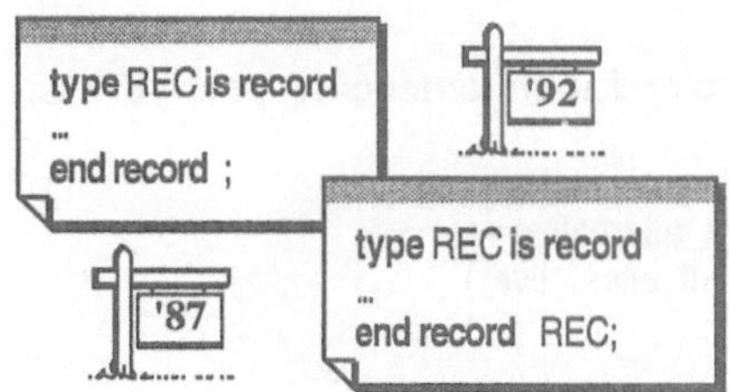

Fig 29.4 Examples of Record Type Declaration Syntax

The exact syntax of record type declaration is

```
record
        element_declaration
        { element_declaration }
end record [ record_type_simple_name ] ;
```

29.2.5. Labeled Sequential Statements Syntax

Since the structured sequential statements **case** and **if** can be labeled in VHDL'92 (cf. chapter 35), it is optionally possible to append this label at the end of the statement.

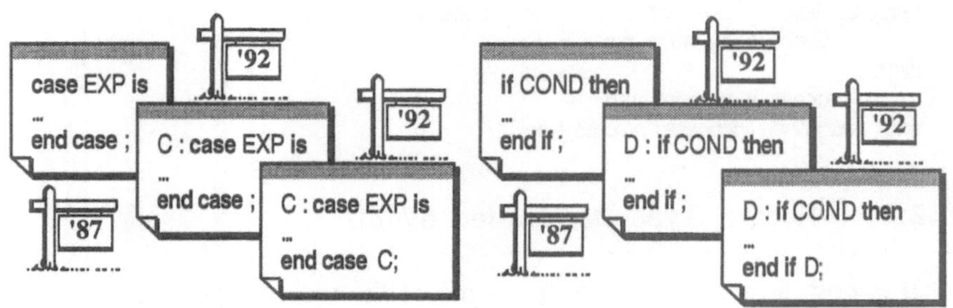

Fig 29.5 Examples of if and case Statement Syntax

The syntax of **case** and **if** statements respectively is:

```
[ case_label : ] case expression is
        case_statement_alternative
        { case_statement_alternative }
end case [ case_label ] ;

[ if_label : ] if condition then
        sequence_of_statements
{ elsif condition then
        sequence_of_statements }
[ else
        sequence_of_statements ]
end if [ if_label ] ;
```

29.2.6. Subprogram Body Syntax

It is now possible to repeat the kind of subprogram (reserved word **procedure** or **function**) after the final **end** (and before the optional name). The general syntax of a subprogram body is

```
subprogram_specification is
        subprogram_declarative_part
begin
        subprogram_statement_part
end [ subprogram_kind ] [ designator ] ;
```

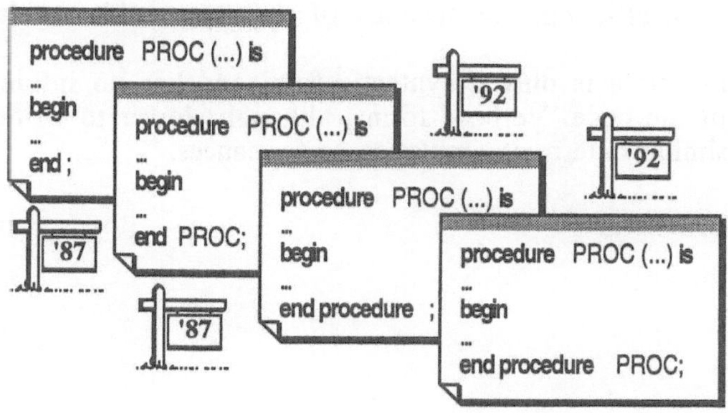

Fig 29.6 Examples of Procedure Body Syntax

29.2.7. Syntax Exceptions

Because **loop** and **generate** are usually read as verbs, it was not consistent to allow the reserved word (but also verb) **is** immediately after. Thus, the **generate**[20] and **loop** statements are as below.

```
generate_label : generation_scheme generate
        [ { block_declarative_item }
begin ]
        { concurrent_statement }
end generate [ generate_label ] ;

[ loop_label : ]   [ iteration_scheme ] loop
        sequence_of_statements
end loop [ loop_label ] ;
```

29.3. USING THE FEATURE

A good practice is to write such features in the most verbose way. In complex sequences of code involving many embedded statements, the indication **end procedure** PROC; is more profitable to the user than the strictly necessary and anonymous **end**;. Such writing is more readable and allows more

[20] The new declarative part that appears in the generate statement syntax is discussed in chapter 19

compilation checks, e.g., consistency of the name with the label of the construct.

Of course, this is only a syntactic form and has no influence on the efficiency of the code. Verbose forms, although slower to write than short forms, are similar in term of simulation performances.

30. STATIC EXPRESSIONS

Warning: this chapter concerns global and local staticness. This subject is very important for the safety and portability of VHDL models, but has very few practical implications for the designer, in the sense that it brings no new functionalities nor enforces a specific style. This chapter will mainly interest implementors and support teams rather than designers.

LRM REFERENCES: 7.4

30.1. BACKGROUND

A VHDL model must be analyzed before execution. During analysis, the compiler performs some checks, such as making sure that all identifiers used in an expression are declared and are of an appropriate type.

If the analysis is successful, the model can be executed. This execution proceeds in three steps:

- First the model is elaborated, that is, all the objects declared in the model (constants, variables, signals, subprograms, etc.) are allocated memory space in the mainframe of the computer. During this step, some other checks are performed on the model.
- Then the initialization phase computes all initial values of drivers and executes all processes until they are suspended.
- Last comes the simulation phase, during which the simulator monitors time advances and executes processes. This execution may cause an update of signals and variables. The simulator must ensure that the values assigned belong to the subtype indication of the object being updated.

30.1.1. Static Expressions

We can see that some checks are performed during analysis of the model, others during elaboration, and the rest during simulation. The first kinds of checks are said to be *locally static*; the second are *globally static*; and the last are *dynamic*.

Furthermore, during analysis, some expressions may be evaluated: they are *locally static expressions*. During elaboration, other expressions may be evaluated: these are *globally static expressions*. The phrase *static expressions* is used to designate either locally static or globally static expressions. An expression that is not static can only be evaluated dynamically during the initialization or simulation phase.

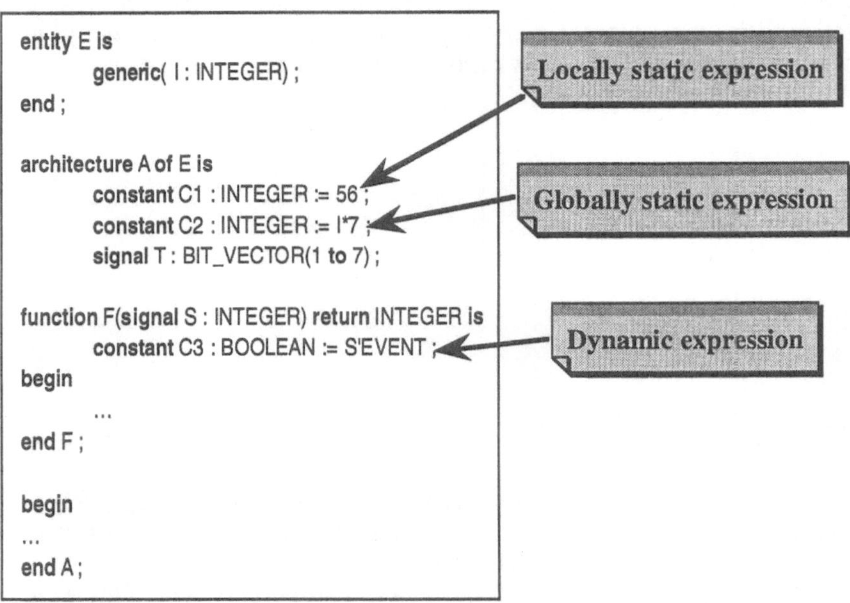

Fig 30.1 Expression Classification

In figure 30.1, constant C1 is initialized with a locally static expression. For constant declaration C2, the expression includes a reference to I for which a value can be known only after it has been constructed in main memory (that is during elaboration); therefore, expression I*7 is globally static.

The value of constant declaration C3 can be known only when function F is executed. This function can be executed only during simulation, because its argument is a signal. Expression S'EVENT is not a static expression; it is a dynamic expression!

30.1.2. Static Names

In a VHDL text, declarations associate identifiers with declared items. For example, given the declarations

```
signal S : INTEGER;
constant C : BIT_VECTOR(1 to 4) := "1010";
```

the identifier S is associated with a signal of type INTEGER, and C with a constant of type BIT_VECTOR with value "1010."

Names in VHDL encompass all the possibilities provided to reference part or all of the declared items. The evaluation of a name determines the declared item. For example, the name S determines the signal previously declared. The name C(3) is an indexed name that evaluates to the subelement at index position 3 of the constant declared above.

One use of names in a VHDL text is to determine objects on which to perform actions. For example, given the declaration:

```
variable V : BIT_VECTOR(1 to 10);
```

the statement

```
V(3):='1';
```

identifies the subelement at index position 3 of variable V to be the target of a variable assignment statement.

If the object denoted by a name can be fully determined during analysis of the VHDL text, the name is said to be **locally static.** For example, V(3) as shown previously is locally static.

A name like V(I), where I is a generic parameter is not locally static because the index position of the item referred to can only be known during elaboration. All such names that can be fully determined during elaboration are **globally static**.

When the name cannot be determined during elaboration, it is a dynamic name. For example, V(N), where N is a signal or variable of type INTEGER, is **dynamic** because the value of N can be evaluated only during execution.

30.2. DESCRIPTION

The concept of static or globally static expressions defined in VHDL is useful to enforce checks and rules such as those requiring the drivers of signals to be created during elaboration: after elaboration, the net is completely constructed, and no further connection (signal) or component instantiation can appear dynamically. Some problems are identified in the VHDL'87 LRM definition of static expressions and names.

30.2.1. Staticness of Physical Literals

30.2.1.1. VHDL'87 Problem

In VHDL'87, physical literals are considered to be locally static. This means that given the VHDL model

```
entity ADDER is
        port (    A,B : INTEGER;
                  S : out INTEGER);
end;

architecture DATAFLOW of ADDER is
begin
        S <= A+B after 2 ns;
end DATAFLOW;
```

the expression *2 ns* is locally static; therefore, a compiler may choose to evaluate it during analysis. But evaluating such an expression (i.e., coding the result) requires knowing the secondary time unit, if any, chosen as the simulation time resolution limit. Furthermore, such a strategy would imply an ability to verify statically (during elaboration) that all the models used to build the design hierarchy do not have conflicting resolution limits.

This resolution limit is the time unit that may be chosen by the user to replace the default fs unit as the smallest TIME unit in case the simulator is not able to fully support all the TIME units.[21]

So if sec is the resolution limit, then 2 ns must evaluate to 0 sec because “any TIME values smaller than this limit are truncated to zero (0)” (LRM 3.1.3.1).

This implies that the resolution limit of the simulator must be specified before any analysis that is in the VHDL text. But the language neither provides

---

[21] This would require at least 64 bits integers, which is not actually required by the LRM'92.

a standard way to specify the resolution limit in a VHDL text, nor requires such a specification in a design unit.

30.2.1.2. VHDL'92 Solution

Physical literals of type STD.STANDARD.TIME are no longer locally static. They are globally static. TIME expressions can only be evaluated at elaboration or later on- that is, when the resolution limit is known.

30.2.2. Staticness of Names Denoting Access Values

30.2.2.1. VHDL'87 Problem

Names designating objects denoted by access values are locally static if they do not contain any expression or if all the expressions they contain are locally static, as shown by the example below. Given the declarations:

```
subtype BIT_VECTOR2 is BIT_VECTOR(0 to 1);
type PTR_BIT_VECTOR2 is access BIT_VECTOR2;
variable A,B : PTR_BIT_VECTOR2;
constant C : BIT_VECTOR2 :="11";
```

A(1) and B(1) denote locally static names, since the only expression they include is the literal 1, which is a locally static expression (VHDL'87 LRM 7.4). The trouble is that it is impossible during analysis to check that they denote different objects. The aggregate variable assignment statement

```
(A(1),B(1)) := C;
```

therefore cannot be analyzed properly as required, although the subelements in the aggregate are denoted by locally static names!

30.2.2.2. VHDL'92 Solution

A locally static name cannot designate either an object or a value of an access type. Therefore the variable target (A(1),B(1)) is not legal and will be rejected during analysis.

30.2.3. Aliases as Static Names

30.2.3.1. VHDL'87 Problem

Another problem is related to names denoting aliases. Consider the declaration

```
signal S : BIT_VECTOR(1 to N+1);
alias X : BIT is S(N);
```

where N is a generic parameter. The name X matches the conditions required for locally static names (it does not contain any expression) and therefore can appear in an aggregate target of a signal assignment statement, whereas S(N) cannot because it is a globally static name!

30.2.3.2. VHDL'92 Solution

A name that is an alias is locally static if and only if it is associated with a locally static name in its declaration. With this new definition, the alias X above is not locally static because it is associated with S(N), which is globally static because N is a generic constant.

30.2.4. Staticness of Subprogram Resident Constants

30.2.4.1. VHDL'87 Problem

If we strictly conform to VHDL'87, a constant declared in a subprogram can be considered as a static expression even if it is dynamically elaborated. This raises some conceptual problems as illustrated in the example below:

```
entity E is
    generic (I : INTEGER);
end E;

architecture ARC of E is
    signal S, T, R, W : BIT_VECTOR (1 to 10);
begin
    process
        variable A : INTEGER;
```

```
            procedure P is
                constant C : INTEGER := 2;
            begin
                S(C) <= '1';
                W(C to C+4) <= "11111";
                ...
            end P;
        begin
            P;
            T(I) <= '1';
            R(A) <= '0';
            ...
        end process;
    end ARC;
```

During elaboration of the previous process, a driver should be created for the signals S, W, T, and R because they may be modified within the process. Drivers serve to hold the modifications performed on signal targets by assignment statements. VHDL rules for signal targets of composite type require that drivers be created only for the subelements that are potentially modified.

The portion potentially affected by an assignment statement is defined to be the **longest static prefix of** the signal because it is evaluated during elaboration. In the example above, T(I) is completely determined during elaboration because I is a generic parameter whose value is completely determined then.

So a driver will be created for the subelement at position I. For R(A), the situation is more difficult, because A is a variable and its value may be updated during simulation. Therefore, no assumption can be made during elaboration as to which subelement will be modified. In this case, the longest static prefix is R - that is a driver will be created for each subelement of R!

The way to determine longest static prefix of signal targets is described in the LRM. The problem stems from the fact that S(C) is considered to be a static prefix although the value of C cannot be known until dynamic elaboration of subprogram P.

The problem is similar for an assignment to a slice W(C **to** C+4) performed within P with the slice expression denoted by a locally static range constraint (C **to** C+4).

30.2.4.2. VHDL'92 Solution

A primary expression denoting objects cannot be globally static if it denotes a dynamically elaborated object. Therefore the constant C declared in procedure P above is no longer considered static, because it is dynamically elaborated during each call to procedure P.

So the signal target S(C) is no more static. Still, constant C is locally static because its subtype (INTEGER) and initial value (18) are completely known during analysis. Thus a locally static expression is not necessarily globally static in VHDL'92. So the longest static prefix of S(C) is S. A driver will be created for each subelement of S, although only S(2) will be potentially affected.

30.2.5. Staticness of Discrete Ranges

30.2.5.1. VHDL'87 Problem

A range describes a set of contiguous discrete values. In VHDL these values may be expressed using either of the two forms:

```
<SIMPLE_EXPRESSION1> to <SIMPLE_EXPRESSION2>
```

For example, 1 **to** 7 refers to the integer values ranging from 1 to 7.

```
<PREFIX>'RANGE
```

For example, given the declarations:

```
subtype POSITIVE is INTEGER range 1 to INTEGER'HIGH;
signal S : BIT_VECTOR(2 to 10);
```

POSITIVE'RANGE refers to the integer values ranging from 1 to INTEGER'HIGH, and S'RANGE to integers ranging from 2 to 10.

VHDL'92 introduces the notion of **locally static range** to designate ranges whose value set can be known during analysis, such as the range

```
1 to 7
```

Globally static range are ranges that can be fully determined during elaboration, e.g.,

```
1 to N
```

where N is a generic parameter.

A static range is either a locally static range or a globally static range. A **dynamic range** is a range that can be determined only during execution of the model, such as

```
1 to N
```

where N is a subprogram parameter.

But there is a problem in the VHDL'87 definition of globally static discrete ranges:

"A globally static range is a range whose bounds are globally static expressions" (LRM 7.4, pages 7-16)

It is obvious that a range such as

```
1 to 8
```

is static. What about ranges expressed using the 'RANGE attribute? Nothing in the VHDL'87 definition allows us to determine the staticness of BIT4'RANGE given the declaration of BIT4 below.

The notion of static range has a direct impact on the determination of the size of the drivers of signal targets expressed in the form of a slice, as in the following:

```
subtype BIT4 is BIT_VECTOR(1 to 4);
signal S : BIT_VECTOR(1 to 8);
S(BIT4'RANGE) <= "1111";
```

What are the bound expressions of BIT4'RANGE? If BIT4'RANGE is considered static (which corresponds to our intuition given the declaration of BIT4), then the name S(BIT4'RANGE) is static, which means that drivers will be created during elaboration only for S subelements at position ranging from 1 to 4.

On the other hand, if BIT4'RANGE is considered not static, then a driver has to be created for each subelement of signal S. The rationale behind such a decision in VHDL is that if the range is not static, it may vary dynamically, which means that no assumption may be made concerning which portions of the signal will be affected: presumably all of them are affected. If the range can be determined statically (during elaboration), then only the portion determined by the range is affected.

30.2.5.2. VHDL'92 Solution

A range expressed in one of the form

```
<simple_expressession1> to <simple_expressession2>
```

is globally (respectively, locally) static if the bounds (<simple_expressession1> and <simple_expressession2>) are globally (respectively, locally) static - for example the range

```
I to I+9
```

where I, a generic parameter, is globally static. If instead, I is a constant declared as

```
constant I : INTEGER := 9;
```

then the range above is locally static. If I is a subprogram parameter, the range is not static.

If the range is expressed in the form

```
<prefix>'RANGE
```

then it is globally (respectively, locally) static if the prefix denotes either a globally (respectively, locally) static subtype or an object having a globally (respectively, locally) static subtype.

Given this new definition of static range, the range BIT4'RANGE is globally static and even locally static; hence, S(BIT4'RANGE) is a locally static signal target (assuming, for example, that BIT4 is a subtype within the same declarative zone as S).

30.2.6. Staticness of Predefined Attributes

30.2.6.1. VHDL'87 Problem

Another problem in VHDL'87 is to determine the staticness of attribute expressions such as S'LEFT in the example:

```
S(S'LEFT)<='1';
```

It is unclear in VHDL'87 whether or not S'LEFT is a static primary. Therefore, the longest static prefix is hard to determine, although in this case it seems clear that the designers want to modify just the subelement at the index position given by S'LEFT.

30.2.6.2. VHDL'92 Solution

Predefined value attributes other than function attributes 'EVENT, 'ACTIVE, 'LAST_EVENT, 'LAST_ACTIVE, 'DRIVING, and 'DRIVING_VALUE, prefixed by globally static or locally static subtypes or objects of such subtypes are, respectively, globally static or locally static.

In case the attribute is a function, the actual parameter (if any) should be a globally static or locally static expression. For example, given the declarations

```
signal S : BIT_VECTOR(0 to 8);
subtype VECT is INTEGER range 1 to N;--N denoting a generic constant
procedure F(A : BIT_VECTOR);
```

the attribute S'LEFT is locally static because S has a locally static subtype. VECT'LEFT, on the other hand, is globally static because VECT is a globally static subtype. Within the body of procedure F, A'LEFT is not static because its subtype is elaborated dynamically during the corresponding procedure call.

30.2.7. Extensions to the Notion of Static Expression

The definition of static expression seems to be overly restrictive, since it disallows indexed names, record aggregates, and function calls. In some cases allowing a function call might prove useful; for instance, a designer may want to initialize a ROM constant by a call to a function that will open a file to read the content of the ROM.

Besides fixing the inconsistencies of VHDL'87, the scope of static expressions is extended to include calls to user-defined pure functions (for the notion of pure functions, see chapter 10), indexed names, record aggregate, and so forth.

30.3. VHDL'87 / VHDL'92

The problems mentioned above were solved by an ISAC issue, and some of them have already been fixed in many available simulators. For example, many simulators on the market already accept user-defined function calls as static expressions.

30.4. USING THE FEATURE

After discussing the subtleties of expression and name staticness, a designer may be tempted to think that these notions have no impact on the VHDL model crafted. This is not true, as illustrated by the ROM example.

One natural way to specify the content of a ROM is to write it in a text file. The first two lines of the file may be used to specify the size of the ROM and the size of the words. For example,

```
4
8
'1' '0' '0' '1' '0' '0' '1' '1'
'0' '0' '0' '1' '1' '1' '1' '0'
'1' '0' '0' '1' '0' '1' '1' '0'
'1' '0' '0' '1' '0' '0' '0' '1'
```

may represent the specification of a ROM containing four bytes. In a given design, a ROM can be represented by a constant declaration, and the file can be used to compute the value of the ROM. To do this, we must declare the type of the ROM and write a function to open the file and get the ROM contents. This is done in the package ROM_PKG:

```
package ROM_PKG is
        type MATRIX is array(NATURAL range <>, NATURAL range <>) of BIT;
        function READ_ROM_CONTENT(FILE_NAME : STRING) return MATRIX;
end ROM_PKG;
```

Since the size of the ROM is specified in the file, its type is an unconstrained array.

Since the ROM is described by a text file, we can use the package TEXTIO to implement the body of ROM_PKG. So the use clause

```
use STD.TEXTIO.all;
```

can be inserted before the package body to give us the possibility of directly using items declared in TEXTIO.

The function READ_ROM must open the file and read the content of the ROM. This content must be stored in a local variable. The trouble is that we cannot declare such a variable unless we know its sizes, which are stored in the file! So the file must be open to get the size. For that, the file must be declared and opened before the variable. Furthermore, we need a function to read the size and width from the file. Assuming the file is already open in READ_MODE, such a function may be written as an impure function.

Coming back to the function READ_ROM_CONTENT, we may write it as follows:

```
function READ_ROM_CONTENT(FILE_NAME : STRING) return MATRIX is
        variable NB_ROW, WIDTH : INTEGER;
        file ROM_FILE : TEXT open READ_MODE is FILE_NAME;
```

```
        impure function GET_SIZE return INTEGER is
          variable L : LINE;
          variable SIZE : INTEGER;
        begin
         READLINE(THE_FILE, L);
         READ(L, SIZE);-- get the size
         return SIZE;
        end function GET_SIZE;

         variable CONTENT : MATRIX(       1 to GET_SIZE,
                                          -- get the ROM size
                                          1 to GET_SIZE);
                                          -- get the size of the words
         variable L : LINE;
begin
  for i in CONTENT'RANGE(1) loop
         READLINE(ROM_FILE, L);
         for j in CONTENT'RANGE(2) loop
              READ(CONTENT(i,  j));
         end loop;
  end loop
  return CONTENT;
end READ_ROM_CONTENT;
```

Whenever a ROM is needed in a model, assuming the appropriate use clause for package ROM_PKG, the following constant declaration is sufficient:

```
constant ROM : MATRIX := READ_ROM_CONTENT(THE_FILE_NAME);
```

where THE_FILE_NAME is the name of the text file describing the ROM.

31. RUN-TIME CHECKS

Warning: this chapter discusses dynamic checks that must be performed at run-time. This subject is very important for the safety and portability of VHDL models, but has very few practical implications for the designer, in the sense that it brings no new functionalities nor enforces a specific style. This chapter will mainly be of interest to implementors and support teams, rather than designers.

LRM REFERENCES: 12

31.1. BACKGROUND

Certain checks required by VHDL cannot be made during analysis or elaboration of the model because all the necessary information is not yet available. For example, consider the declaration:

```
signal S ,T : POSITIVE;
```

The signal assignment

```
S <=T-8;
```

is valid only if the expression T-8 belongs to the subtype of S that is POSITIVE. Such a check cannot be performed statically: it must be verified dynamically, that is, during execution of the model after the effective value of

T is known. All checks that can only be performed during execution of the model are called run-time checks.

31.2. DESCRIPTION

In the VHDL'87 LRM, the set of run-time checks to be enforced by a conforming implementation is not clearly and completely specified. This is damaging to the portability of VHDL models, because different implementations might not perform the same set of dynamic checks. A description of subprogram parameter run-time checks best illustrates this problem.

31.2.1. Run-Time Check of Non-Signal Parameters

31.2.1.1. VHDL'87 Inconsistency

Section 8.5 of the VHDL'87 LRM specifies the following run-time checks on non-signal subprogram parameters:

- before subprogram execution, all the arguments of the subprogram are checked for compatibility with the corresponding formal parameter;
- for the **in** or **inout** parameter, the value of the actual is copied to the corresponding formal;
- after execution, the value of **out** or **inout** formal parameters is copied back to the corresponding actual;
- no further check is explicitly required to ensure that the values assigned to the actuals belong to their subtype indication.

Consider the subprogram

```
procedure P(    variable A : inout INTEGER;
                constant  B : in INTEGER;
                variable C : out INTEGER) is
begin
   A := A+B;
   C := A-8;
end P;
```

and the declarations

```
variable T : INTEGER range 1 to 20 :=1;
variable U : INTEGER range 1 to 9;
```

A call such as

```
P(T, -5, U);
```

satisfies all the dynamic checks expressed previously: T,-5, and U are of **type** INTEGER, as required for formal parameters A, B, and C. So the execution of such a call should not produce an error. But the execution of that procedure call is equivalent to the execution of the sequence

```
A := T;
B := -5;
A := A+B;
C := A-8;
T := A;
U := C;
```

which assigns the value -4 to T and -12 to C. These values violate the constraints of their subtype indication: the value of T should lie between 1 and 20 and value of U between 1 and 9.

31.2.1.2. VHDL'92 Solution

Dynamic checks of subprogram scalar parameters are modified by VHDL'92 in the following way:

- a dynamic check must be made to ensure that the values of actuals copied into corresponding **in** or **inout** formal parameters belong to the parameter subtype indication (same as VHDL'87); and furthermore,
- before copying the value of **out** or **inout** formal parameters, a check must ensure that the values copied back to actual parameters belong to the subtype indication of the actuals. So in the above example, such a check would fail because -4 is out of the range of the actual T (which is 1 **to** 20) and -12 is out of the range of the actual U (which is 1 **to** 9).

These additional run-time checks are especially useful when function conversion or type conversion is used, because VHDL'87 did not explicitly require the returned result to belong to the subtype indication of the corresponding formal or actual.

31.2.2. Run-Time Checks of Signal Parameters

In the case of signal parameters, the rule is that all actions performed on a signal formal be in fact equivalent to directly applying the same actions on the effective signal given as actual parameter. So, in almost all of the cases, the formal signal declaration is viewed as an alias of the actual signal.

Given the declarations

```
subtype INTEGER9 is INTEGER range 1 to 9;

procedure ASSIGN(signal S : out INTEGER9; N : INTEGER) is
begin
        S <= N;
end;

signal SIG : INTEGER;
```

the execution of the subprogram call

```
ASSIGN(SIG,18);
```

is equivalent to executing

```
SIG <= 18;
```

By the time the signal assignment is executed, no subtype check is required to ensure that, for example, the value assigned to the formal signal parameter of mode **out** or **inout** belongs to its subtype, since the assignment is supposed to be directly performed on the actual. LRM 12.1 states that "for scalar signal S, both driving and effective values must belong to the subtype of the signal." The subtype indication seems useless. So in the example above, the procedure call could be executed without any error.

31.2.2.1. VHDL'92 Solution

VHDL'92 add the following constraints to signals associated with formal signal parameters in subprogram call: for association of signals with scalar formal signal parameters, the subtype indication of the actual signals must have the same bounds and direction as the formal parameter's subtype indication. With this rule, the call

```
ASSIGN(SIG,18);
```

will be erroneous, because the bounds of SIG (INTEGER'LOW and INTEGER'HIGH) are not the same as the bounds of the formal parameter S (1 and 9).

31.2.3. VHDL'87/VHDL'92

The workaround for the issues raised above is to make sure when scalar parameters are specified in subprograms that the actuals have the same subtype constraints. Another way is to declare scalar parameters without any constraint and to use assert statements within subprogram bodies to enforce the intended constraint check. The ASSIGN subprogram could be rewritten as

```
procedure ASSIGN(signal S : out INTEGER; N : INTEGER) is
begin
        assert ((N <= 1) and (N >= 9)); --checks that value (N) assigned to S lies between 1 and 9
        S <= N;
end;
```

The new requirements added by VHDL'92 will enhance the portability of VHDL models across different platforms, but many models may need to be modified to comply with the new rules. So the changes are not upward compatible.

Many other run-time checks are made explicit in VHDL'92 but they have very little impact on the code written by the designers and therefore are not described here.

32. INTERFACE LIST

LRM REFERENCES: 4.3.2.1

32.1. BACKGROUND

In VHDL, generic parameters, port declarations, and subprogram parameters are described by interface lists. It is not really clear in VHDL'87 whether or not a given formal can be referenced in the declaration of another formal when both appear in the same interface list. To illustrate this, consider the specification of an adder. It may be tempting to write it as

```
entity GENERAL_ADDER is
      port(    A : BIT_VECTOR; B : BIT_VECTOR(A'RANGE);
               S : BIT_VECTOR(0 to A'LENGTH));
end GENERAL_ADDER;
```

or, if a concurrent procedure call is preferred, as

```
procedure GENERAL_ADDER(     signal A : BIT_VECTOR;
                             signal B : BIT_VECTOR(A'RANGE);
                             signal S : out BIT_VECTOR(0 to A'LENGTH));
```

In both cases, the intent of the designer seems clear: the adder is very general and depends solely on the size of A. B is declared with the same range, and S is declared with an extra BIT for the carry out. The trouble is that B and

S cannot be fully determined before A, which contradicts the VHDL'87 rule that they can be elaborated (i.e. fully determined) in any order!

32.2. DESCRIPTION

VHDL'92 brings some clarification and disallows the declaration of a given formal to reference another formal declared in the same interface list. So the examples above are explicitly illegal in VHDL'92.

32.3. USING THE FEATURE

Coming back to the specification of our adder, we could write it as

```
entity GENERAL_ADDER is
        generic( N : POSITIVE);
        port(    A : BIT_VECTOR(0 to N-1);
                 B : BIT_VECTOR(0 to N-1);
                 S : BIT_VECTOR(0 to N));
end GENERAL_ADDER;
```

or, using a subprogram, as

```
procedure GENERAL_ADDER(    signal A : BIT_VECTOR;
                            signal B : BIT_VECTOR;
                            signal S : out BIT_VECTOR);
```

Notice the reference to the generic parameter N within the port declaration: this is allowed because generic and port declarations are distinct interface lists. In the case of subprograms, there is no separate generic parameter declaration so the only choice is to impose no constraint at all on its parameter declaration. In any case the checks can be done (at run-time) in the subprogram body with the following assert statement:

```
assert((A'LENGTH=B'LENGTH)and(A'LENGTH+1=S'LENGTH))
        report "Error in GENERAL_ADDER : check parameter sizes" severity ERROR;
```

33. ASSOCIATION LIST

LRM REFERENCES: 4.3.2.1

33.1. BACKGROUND

Association lists are optionally used in VHDL to assign signals to ports and to specify actuals to subprogram or generic parameters. In VHDL, it is possible to connect signals to ports having different types using type conversion function for example. Consider the following declarations:

```
subtype WORD4 is BIT_VECTOR(0 to 3);
function TO_INTEGER(A : BIT_VECTOR ) return INTEGER;
function TO_WORD4(ARG : INTEGER) return WORD4;
signal INPUT1,INPUT2,OUTPUT : INTEGER;
component ADDER4
        port (A,B : WORD4; S : BIT_VECTOR(0 to 4));
end component ADDER4;
```

Component ADDER4 can be instantiated using signals INPUT1, INPUT2, and OUTPUT. But since their type is not compatible with the ADDER4 ports type, type conversion functions must be used. This can be done as follows:

```
INSTANCE_ADDER4 : ADDER4 port map(      A => TO_WORD4(INPUT1),
                                        B => TO_WORD4(INPUT2),
                                        TO_INTEGER(S) => OUTPUT);
```

Notice that since information flows from INPUT1 and INPUT2 to the ports A and B, the conversion function that is used transforms values of the formers to values compatible with the latters. For S it is the contrary: S is an output port -that is information flows from S to OUTPUT- so the conversion function transforms the value of S into an integer. If the information flows in both directions as is the case for ports of mode **inout**, then the conversion must be done in both directions. Assuming T is an **inout** port of subtype WORD4 to be connected to signal S_INT of type integer, we would have the association:

```
TO_INTEGER(T) => TO_WORD4(S_INT)
```

But what happens if the previous instantiation is written as follows:

```
INSTANCE_ADDER4 : ADDER4 port map( TO_INTEGER(A) => TO_WORD4(INPUT1),
                                   TO_INTEGER(B) => TO_WORD4(INPUT2),
                                   TO_INTEGER(S) => OUTPUT);
```

Here, function conversion TO_INTEGER is useless in the first two associations because information cannot flow from A to INPUT1 or from B to INPUT2, since A and B are in ports. In any case, VHDL'87 allowed such associations to be written, which is nonsense.

33.2. DESCRIPTION

In VHDL'92, function conversion or type conversion can only be used when necessary according to the direction of information flow. So the last instantiation statement is illegal in VHDL'92. The same limitation applies for association lists of configuration specifications or subprogram calls where function conversion or type conversion can be used.

34. RESOLVED SUBELEMENTS IN COMPOSITES

LRM REFERENCES: 4.3.2.2

34.1. BACKGROUND

Signals in VHDL model wires and support propagation of waveforms and interconnection of gates. In fact a concurrent signal assignment statement such as

```
S <= A and B;
```

is equivalent to the gate

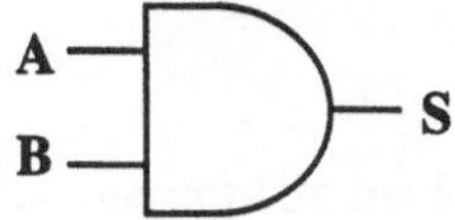

The waveform flowing through S is produced by the and gate connecting A and B. When there are many concurrent assignments to the same signal, the signal is driven by many different sources. In VHDL a signal having many sources must be resolved, that is, associated with a function that will be used by the simulator to compute the value of the signal from the values driven by the sources. Such a function is called a resolution function: it takes an unconstrained array as a parameter and returns a value belonging to the type of the signal. This function must be provided by the designer. For example, the function

```
function WIRED_AND(ARG : BIT_VECTOR) return BIT;
```

can be used as a resolution function in one of two ways:

```
signal S : WIRED_AND BIT;
subtype BIT_RESOLVE is WIRED_AND BIT;
signal S2 : BIT_RESOLVE;
```

The first way is to declare signal S as a resolved signal: association with the resolution function is made in the signal declaration.

The second way is to declare a resolved subtype by associating a resolution function with a type. Notice that BIT_RESOLVE is a subtype of BIT, which means that all operations defined for BIT also apply to BIT_RESOLVE. Furthermore, BIT and BIT_RESOLVE share exactly the same set of values. The only difference is that signals of type BIT may not be assigned in many concurrent statements while BIT_RESOLVE signals may have many sources because of the resolution function associated with BIT_RESOLVE.

34.2. DESCRIPTION

In VHDL, subelements of composite signals (array or record) are viewed as signals of their own and may be resolved in a different way with respect to each other. The type of such composite signals must specify the resolution functions associated with the corresponding subelements. For example, in the record type

```
type COMPLEX is record
        FIRST_PART : BIT_RESOLVE;
        SECOND_PART : BIT;
end record;
```

the first subelement is resolved and the second is not.

In the case of scalar types, VHDL'87 allows the definition of a single scalar type from which many resolved subtypes can be defined that share the same set of values and operations but differ only in the associated resolution. This is also true for composite types taken as a whole. But in this case, the atomicity of scalar subelements is lost: all the subelements of a resolved composite subtype are processed by the same resolution function.

Given a composite type, there is no way in VHDL'87 to derive a subtype that associates resolution functions with the scalar subelements of the composite type. This is because subelement resolution functions, if any, must be specified in the type declaration itself. So the only way to have identically structured objects differing only in the way their subelements are resolved is to declare two different types. But VHDL is a strongly typed language, which means that objects belonging to different types, though sharing the same structure, have different sets of values and thus cannot be freely assigned (value) to each other. Typical examples of such situations may happen in a design where busses with subelements resolved differently need to communicate. For example, consider the declarations below.

```
subtype RESOLVED_AND_BIT is  WIRED_AND BIT;
subtype RESOLVED_OR_BIT is  WIRED_OR BIT;
type RESOLVED_AND_BIT_VECTOR is
                                array (POSITIVE range <>) of  RESOLVED_AND_BIT;
type RESOLVED_OR_BIT_VECTOR is
                                array (POSITIVE range <>) of  RESOLVED_OR_BIT;
type RECORD1_TYPE is record
        A : WIRED_AND_BIT;
        B : WIRED_OR_BIT;
end record;
type RECORD2_TYPE is record
        A : BIT;
        B : WIRED_AND_BIT;
end record;
signal WIRED_OR_SIGNAL : RESOLVED_OR_BIT_VECTOR (1 to 7);
signal WIRED_AND_SIGNAL : RESOLVED_AND_BIT_VECTOR (1 to 7);
signal REC1_SIGNAL : RECORD1_TYPE;
signal REC2_SIGNAL : RECORD2_TYPE;
```

We see that WIRED_OR_SIGNAL and WIRED_AND_SIGNAL have exactly the same structure, that is, they are both arrays of BIT; they differ only in the way their subelements are resolved. In fact this is the reason why two different type declarations are needed! But because they belong to different types, assignments such as

```
WIRED_OR_SIGNAL <= WIRED_AND_SIGNAL;
```

are not allowed. Furthermore, a component instantiation cannot map a port of type RESOLVED_OR_BIT_VECTOR to a signal of type RESOLVED_AND_BIT_VECTOR for the same reason.

The same restrictions apply to signals REC1_SIGNAL and REC2_SIGNAL, for the same reasons.

VHDL'87, through the mechanisms of type conversion and conversion functions, provides a partial solution to the problems stated earlier. In VHDL, array types defining identically structured objects differing only in their resolvedness are compatible types. So objects of one type can be converted into objects of the other type. In VHDL'87, we can thus have

```
WIRED_OR_SIGNAL <= RESOLVED_OR_BIT_VECTOR (WIRED_AND_SIGNAL);
```

In port map associations, type conversion functions are used to solve the problem of type mismatch. But they do not cover all possible situations, especially when the return type of the type-conversion function may not be constrained. This arises when the formal port is of mode **in, inout**, or **linkage** and the actual includes a conversion function.

34.2.1. VHDL'92 Solution

VHDL addresses the problem exposed above by allowing a type conversion in the specification of formal and actual parts of element association. A formal part in an element association of a list association is specified as

```
FORMAL_PART ::=     FORMAL_DESIGNATOR
                    | FUNCTION_NAME ( FORMAL_DESIGNATOR )
                    | TYPE_MARK ( FORMAL_DESIGNATOR )
```

Likewise, an actual part is specified as

```
ACTUAL_PART ::=     ACTUAL_DESIGNATOR
                    | FUNCTION_NAME ( ACTUAL_DESIGNATOR )
                    | TYPE_MARK ( ACTUAL_DESIGNATOR )
```

This feature enhances the possibility allowed in VHDL'87 by type conversion functions; however does not address the case of record types, which must be solved using type conversion functions because no type conversion is allowed for record types. VHDL'92 does not define such a thing as structurally equivalent or as closely related record types!

34.3. VHDL'87/VHDL'92

The semantics of allowing type conversion in association lists is defined in terms of the equivalent type conversion function.

Let us consider the declarations

```
component CIRCUIT is
        port(A : inout WIRED_OR_BIT_VECTOR);
end component CIRCUIT;
```

The semantics of

```
C : CIRCUIT port map ( WIRED_AND_BIT_VECTOR(A) =>
                                WIRED_OR_BIT_VECTOR(WIRED_AND_SIGNAL));
```

is the following piece of code:

```
subtype WIRED_AND_SUBTYPE is WIRED_AND_BIT_VECTOR(1 to 7);
subtype WIRED_OR_SUBTYPE is  WIRED_OR_BIT_VECTOR(1 to 7);

function TO_WIRED_AND_SUBTYPE(A : WIRED_OR_SUBTYPE)
                                        return WIRED_AND_SUBTYPE is
begin
        return WIRED_AND_SUBTYPE(A);
end TO_WIRED_AND_SUBTYPE;

function TO_WIRED_OR_SUBTYPE(A : WIRED_AND_SUBTYPE)
                                        return WIRED_OR_SUBTYPE is
begin
        return WIRED_OR_SUBTYPE(A);
end TO_WIRED_OR_SUBTYPE;
...
begin
...
C : CIRCUIT port map ( TO_WIRED_AND_SUBTYPE(A)=>
                                TO_WIRED_OR_SUBTYPE(WIRED_AND_SIGNAL));
```

34.4. USING THE FEATURE

In all situations where we need to connect composite signals of different closely related types, it is no longer necessary to write functions to perform the conversion. In VHDL'92, it is sufficient to use the type conversion, as illustrated in the previous section.

35. LABELS & USER-DEFINED ATTRIBUTES

LRM REFERENCES: APP A, 4.3.3

35.1. BACKGROUND

The attributing mechanism of VHDL'87 had some flaws:

- It is not possible in VHDL'87 to attribute, for example, a sequential assignment statement or an IF statement, because they have no label attached. It happens that synthesis tools could take advantage of such a facility (by attaching resources or timings to a given portion of VHDL), and that the workaround is rather nasty (pseudo-comments).
- A package body cannot be attributed, because no attribute specification can be written in it.
- Overloaded subprograms cannot be easily attributed, because only the name of the subprogram can be used in the attribute specification.
- Issues had to be resolved concerning the visibility of attributes across primary and secondary units.

35.2. DESCRIPTION

• In VHDL'92, any statement can hold a label. This increases the regularity of the language and provides a hook to synthesis tools: labels can be attributed.

```
process
begin
        SEL1 : if S1 = '1' then
            AFF : S2 <= '0' after 10 ns;
        end if;
end process;
```

• Attributes can be specified from within package bodies. For example:

```
package body MATH is
        attribute DEBUGGED of MATH : package is FALSE;
        ...
end MATH;
```

• Attributes can now be specified using the signature explained in chapter 23:

```
attribute XXX of
            A_FUNCTION[BOOLEAN, CHARACTER return INTEGER] : function is VVV;
```

• Attributes can be specified on enumeration literals, using if necessary the above-mentioned parameter result type profile: this is easily understood by remembering that enumeration literals can be seen as parameterless functions.

```
type ENUM is (...,ELEM,...);
--declaration of ELEM is equivalent -in this respect- to :
--function ELEM return ENUM;
```

• Physical units can also be attributed.

Except for interface objects (the ports), whose attribute specifications extend to the architecture, the attributing is made on a per-declarative part basis, which means that an attribute specification in an entity does not extend to the architecture (see figure 35.1).

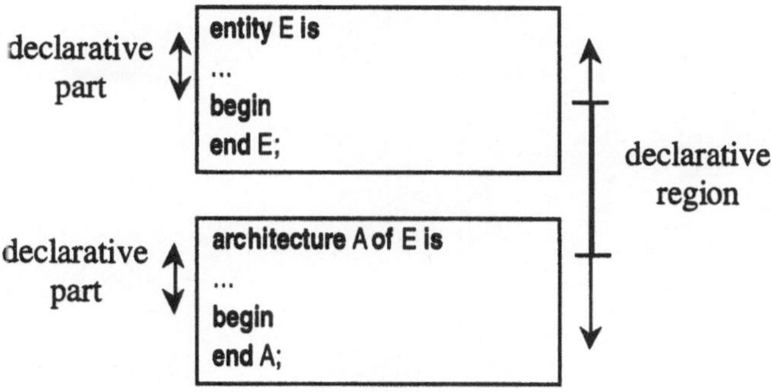

Fig 35.1 Difference Between Declarative Part & Declarative Region

35.3. USING THE FEATURE

The use of attributes -even user-defined attributes- is very implementation dependent. Actually, most of them are in fact implementation defined (typically in synthesis applications), and the reader should carefully check the vendor's documentation in this regard.

The use of attributes in simulation is quite marginal, since they are just constants attached by a syntactic glue to some objects of the language. However, the use of attributes like AUTHOR, REVISION, DEBUGGED, etc., may be contractual or enforced by a local policy.

36. MISCELLANEOUS

LRM REFERENCES: 14 12.1 10.3

36.1. NO NON-LANGUAGE DEFINED UNITS IN STD

In VHDL'87, it was specified that the library STD should contain the package STANDARD (which declares BOOLEAN, INTEGER, etc.) and the package TEXTIO (all human-readable input/outputs).

However, it was not stated that STD should be a "special library," and therefore some implementations had extra units in it - for example, a multi-valued logic package, another IO package, etc.

VHDL'92 specifies that STD must contain

- package STANDARD
- package TEXTIO

and is not allowed to contain anything else.

This change is actually minor and might fix some visibility problems in porting pathological models.

36.2. GENERICS IN ROOT

The language did not specify in LRM'87 if the root entity of a design could have ports and generics.

It is now specified that it may (LRM [12.1]). This means that any implementation is allowed to put any arbitrary restriction on the existence and nature of ports and generics of the root entity.

Consequently, for a design that is supposed to be delivered and simulated on another platform, the only safe methodology is to forbid ports and generics from root entities:

```
entity ROOT is
        -- no generics
        -- no ports
        -- maybe types, constants, statements, etc.
end ROOT;
```

36.3. NEW TIME SUBTYPE

The package STANDARD will now include (LRM [14]):

subtype DELAY_LENGTH **is** TIME **range** 0 FS **to** TIME'HIGH;

This type will be usable in all places where a non-negative time is needed, i.e., nearly everywhere. The use of a named subtype allows for more conciseness, but also gives more chances for optimization. In addition, it may be used by the LRM to make definitions clearer.

36.4. CLARIFICATION OF SOME NAMES

Although many parts of the LRM'87 are dedicated to making names unambiguous, (for example, the overload resolution), some particular cases were overlooked.

36.4.1. Selected Against Expanded

The expanded name has priority over the selected name (LRM [10.3]). This has very little impact on existing designs, because the issue was already known and was fixed in the same way by all implementations:

```
type T is record
        A : INTEGER;
end record;

function F  return T is
        variable A : INTEGER;
        variable I : INTEGER;
begin
        ...
        I := F.A;   -- the variable A in F,
                    -- not the field A of the returned value of a recursive call to F
        ...
end F;
```

36.4.2. Using The Feature

This example shows that the overload resolution can apply even if the overloaded name is a prefix:

```
type FIRST is record
        FIELD : INTEGER;
end record;
type SECOND is record
        FIELD : CHARACTER;
end record;
function F return FIRST;
function F return SECOND;
...
if (F.FIELD = 3)  then        -- from FIRST
...
if (F.FIELD = 'A') then       -- from second
...
```

This second example illustrates that an ambiguity can be found between a function call and the indexing of an array. Let us assume a utility function that is able to return a variable number of spaces:

```
function BLANKS(N : NATURAL :=5) return STRING;
```

This function is to be used, for example, like this:

```
WRITE_MSG(PREFIX & BLANKS(3) & POSTFIX);
```

The problem in VHDL'87 is that it does not compile: the compiler does not know if BLANKS(3) means actually "   " (a string of three characters) or the character ' ' (one space) that is in the third position of the string "   " (five spaces) returned by the call to BLANKS with a default value.

This is again an instance of problems caused by the use of the parenthesis for many purposes, here for indexing an array and for the argument list.

More precisely, there is an ambiguity between BLANKS(3) and BLANKS*(5)*(3) where the (5) is taken from the default. Here the context does not supply the missing information, since "&" is overloaded on strings and characters, but in any case the LRM'87 did not clearly state that the context could make such a case unambiguous.

The LRM'92 will now state that the context may be used to distinguish between selected and indexed names (which, by the way, makes our example definitively ambiguous). But

```
C := BLANCS(4);
```

is legal if C is a character: it will be actually BLANCS(5)(4).

36.5. PRESENTATION OF THE LRM

This is actually not a language change, but rather a change in the way it is documented. The LRM includes a chapter 0 stating its purpose, an Index, Cross References, and an Appendix identifying potentially non-portable constructs.

37. LIST OF RESERVED WORDS

Reserved words in bold are new in VHDL'92. If they have been used as identifiers in VHDL'87 source code, upward compatibility is not ensured and errors will occur during compilation. See chapter 22 to make the corrections.

abs
access
after
alias
all
and
architecture
array
assert
attribute
begin
block
body
buffer
bus
case
component
configuration
constant
disconnect
downto
else
elsif
end
entity
exit
file
for
function
generate
generic
group
guarded
if
impure
in
inertial
inout
is
label
library
linkage
literal
loop
map
mod
nand
new
next
nor
not
null
of
on
open
or
others
out
package
port
postponed
procedure
process
pure
range
record
register
reject
rem
report
return
rol
ror
select
severity
signal
shared
sla
sll
sra
srl
subtype
then
to
transport
type
unaffected
units
until
use
variable
wait
when
while
with
xnor
xor

38. INFORMAL GLOSSARY

VHDL'92 enriches the classical VHDL vocabulary. The goal of this chapter is to give some clues to understand these concepts. The description of each concept is short and very informal, but refers to the part of the book where this notion is described in detail.

• **Direct Instantiation:** A shortcut to the previous sequence: component-declaration, instantiation and configuration. This allows the direct instantiation of an entity in the statement part of an architecture. (see chapter 5)

• **Driving Value:** The contribution of a driver to a signal (as opposed to the *effective value*). For example, in the statement: S <= S1, the effective value of the signal S1 will be a driving value (a contribution) for the signal S. Possibly several driving values will contribute, via the resolution function, to the resolved value of S: the effective value. The attribute 'DRIVING_VALUE returns the driving value of the signal that prefixes it.(see chapter 13)

• **Dynamic Expression:** An expression known only at execution time. For example the expression (S + 3) is dynamic if S is a signal or a variable. (see chapter 30)

• **Effective Value:** The actual value of a signal, once resolved (as opposed to the *driving value*). This value is that which is actually read when the signal is involved in an expression. (see LRM [12.6.1])

• **Foreign:** A subprogram or an architecture that is declared in VHDL but the body of which is possibly written in another language and linked to the declaration by an implementation-dependent mechanism. (see chapter 8)

• **Globally Static (Expression or Name):** An expression or a name that is known at least at elaboration time. For example,

```
constant C1 : INTEGER := C2 + C3;
```

is globally static if C2 and C3 are imported from other packages. (see chapter 30)

• **Group:** A set of VHDL objects designated by their name and grouped under a new name, which can be in turn grouped or attributed (this feature is synthesis oriented). (see chapter 7)

• **Impure Function:** A function that is able to have side effects; i.e., to access objects outside of its own declarative region. Such a function may return different values if called with the same set of arguments. (see chapter 10)

• **Incremental Binding:** A way to configure a design in several steps. Ports (and generics) need not be all fixed at the same level. Generics can also be redefined from an upper level. This mechanism is useful for backannotation. (see chapter 6)

• **Locally Static (Expression or Name):** An expression or a name that is known at compile time. For example, **constant** C1 : INTEGER := 2; is locally static. (see chapter 30)

• **Postponed Process:** A process that is activated when there is nothing else left to do at a given time point (i.e., just before advancing the simulation time). Of course the action of these processes is limited in that they cannot create any new activity at that time point (cannot assign to signals with no delay). Assertions and assignments can also be postponed. (see chapter 3)

• **Pure Function:** A function that complies with the original VHDL'87 definition, i.e. one that can only read and assign to local variables and can only read external constants and its arguments. Such a function is a pure mathematical function in the sense that it will always return the same value for a given set of arguments (a minor exception applies if files, or the function NOW are used). (see chapter 10)

• **Pulse Rejection Limit:** A time that, when given, is used by the inertial delay mechanism. When not given, this time is taken from the after clause of the first waveform element. (see chapter 18)

• **Shared Variable:** A variable that is declared in a concurrent environment (architecture declarative part, or package, etc.). It can be read and written by <u>different</u> processes many times during the <u>same</u> simulation cycle; its value is thus dependent on the order of execution of the processes. This introduces non-determinism in a simulation, and is useful for system-level design. (see chapter 4)

• **Signature:** Applies to subprograms and enumeration literals (seen here as functions without parameters). A signature defines the number and types of arguments and return type, and typically allows for an unambiguous specification for use clauses and aliases in case of overloading. (see chapter 23)

39. INDEX